AUG 3 0 1976

Date Due

AUG 1 0 1976			
SEP 8 1976			
OCT 6 1976			
JUN 2 1 1978			
JUL 6 1978			
FEB 0 5 1980			

TECHNICAL INFORMATION LIBRARIES

MURRAY HILL

164492   MH   01   1974
INT SYMP ARCHITECTURAL A
COUSTICS
    725.8/I61   JUL 27 1976

BELL TELEPHONE LABORATORIES

BRODART   PRINTED IN U.S.A.

AUDITORIUM ACOUSTICS

*The Proceedings of an International Symposium on Architectural Acoustics
held at Heriot-Watt University, Edinburgh, Scotland*

AUDITORIUM ACOUSTICS

Edited by

ROBIN MACKENZIE

Lecturer in Architectural Acoustics, Heriot-Watt University, Edinburgh, Scotland

A HALSTED PRESS BOOK

JOHN WILEY & SONS
NEW YORK—TORONTO

PUBLISHED IN THE U.S.A. AND CANADA BY
HALSTED PRESS
A DIVISION OF JOHN WILEY & SONS, INC., NEW YORK

Library of Congress Cataloging in Publication Data
International Symposium on Architectural Acoustics,
   Heriot-Watt University, 1974.
   Auditorium acoustics.

   "A Halsted Press book."
   Includes index.
     1. Architectural acoustics—Congresses. I.
Mackenzie, Robin. II. Title.
NA2800.156 1974 729′.29   75-16445
ISBN 0-470-56284-6

WITH 12 TABLES AND 162 ILLUSTRATIONS

© APPLIED SCIENCE PUBLISHERS LTD 1975

All rights reserved. No part of this publication may be reproduced, stored in a retrieval system or transmitted in any form or by any means, electronic, mechanical, photocopying, recording, or otherwise, without the prior written permission of the publishers, Applied Science Publishers Ltd, Ripple Road, Barking, Essex, England

Printed in Great Britain at the Alden Press, Oxford

Dedicated to the memory of
Vern Oliver Knudsen

List of Contributors

RICHARD H. BOLT
Bolt, Beranek and Newman Inc., Cambridge, Massachusetts, U.S.A.

SANDY BROWN
Sandy Brown Associates, Conway Street, London, England.

ALEXANDER N. BURD
Sandy Brown Associates, Conway Street, London, England.

LOTHAR CREMER
Institut für Technische Akustik, Technische Universität, Berlin, Federal Republic of Germany.

B. DAY
Department of Architecture, University of Bristol, Bristol, England.

R. GERLACH
III Physikalisches Institut der Universität, Göttingen, Federal Republic of Germany.

VILHELM LASSEN JORDAN
Gevninge, DK 4000 Roskilde, Denmark.

R. LAWRENCE KIRKEGAARD
Bolt, Beranek and Newman Inc., Downers Grove, Illinois, U.S.A.

HEINRICH KUTTRUFF
Institut für Technische Akustik der Rheinisch-Westfalischen Technischen Hochschule, Aachen, Federal Republic of Germany.

ZYUN-ITI MAEKAWA
Faculty of Engineering, Kobe University, Rokko Nada Kobe 657, Japan.

SIR ROBERT MATTHEW
Robert Matthew, Johnson-Marshall & Partners, Edinburgh, Scotland.

PAUL NEWMAN
Department of Architecture, Heriot-Watt University, Edinburgh, Scotland.

PETER H. PARKIN
Building Research Establishment, Garston, Watford, England.

G. PLENGE
Heinrich Hertz Institut, Berlin-Charlottenburg, Federal Republic of Germany.

THEODORE J. SCHULTZ
Bolt, Beranek and Newman Inc., Waltham, Massachusetts, U.S.A.

MANFRED R. SCHROEDER
III Physikalisches Institut der Universität, Göttingen, Federal Republic of Germany.

R. W. B. STEPHENS
President, of the Institute of Acoustics, England.

V. B. TORRANCE
Department of Building, Heriot-Watt University, Edinburgh, Scotland.

PAUL S. VENEKLASEN
Paul S. Veneklasen & Associates, Santa Monica, California, U.S.A.

H. WILKENS
Heinrich Hertz Institut, Berlin-Charlottenburg, Federal Republic of Germany.

Preface

The seventeen papers given in this book were presented by invitation at the International Symposium on Architectural Acoustics held at the Heriot-Watt University, Edinburgh.

During the early stages of organising the conference it was proposed that the papers should cover a broad field within the realms of Room Acoustics. It soon became apparent however, that there was sufficient demand for a symposium specifically on the subject of Auditorium Acoustics. The conference consisted of an introductory session followed by four technical sessions. The order in which the papers appear in this text is similar to that in which they were presented. During the introductory session review papers were presented by two of the most eminent speakers in the field of Architecture and Acoustics respectively, Sir Robert Matthew and Dr. Richard Bolt.

In the first technical section which is based on General Design the first of the two papers covers design considerations from the viewpoint of the Acoustic Consultant whilst the second paper refers to the acoustic qualities associated with some of the finest halls in the world and gives an analysis of the more important details in their design.

The next section relates to the use of models in acoustic research and design. The first paper describes the scale model measurements involved in the design of the Sydney Opera House whilst the remaining papers in this section cover the use of modelling in both research and design work.

In the third technical section a tribute to the memory of Vern Oliver Knudsen is given by Dr. Raymond Stephens, President of the Institute of Acoustics, U.K. Professor Knudsen, one of the leading figures in the field of Architectural Acoustics was due to present a paper at the conference, but died shortly before the meeting. The three remaining papers in this section deal with specific types of auditoria including the concert hall, the multi-purpose hall and a modern concept for the amphitheatre.

In the final section the papers refer to contemporary research and include reference to artificial or 'assisted' forms of improving the acoustics of the auditorium by electronic means. The papers in this section also refer to design by computer simulation, research into reflections from wall surfaces and studies using subjective 'dummy head' techniques.

I wish to thank the speakers for their valuable time and effort spent in the preparation of these papers.

R.K.M.

Contents

LIST OF CONTRIBUTORS vii
PREFACE ix

1. OPENING ADDRESS
   V. B. Torrance 1

2. ACOUSTICS IN PERSPECTIVE
   Sir Robert Matthew 3

3. ARCHITECTURAL ACOUSTICS—A REVIEW OF THIS CENTURY
   Richard H. Bolt 13

General Design

4. DESIGN CONSIDERATIONS FROM THE VIEWPOINT OF THE PROFESSIONAL CONSULTANT
   Paul S. Veneklasen 21

5. MULTI-PURPOSE AUDITORIA: AN AMERICAN PHENOMENON
   Theodore J. Schultz and R. Lawrence Kirkegaard 43

Use of Models

6. MODEL STUDIES WITH PARTICULAR REFERENCE TO THE SYDNEY OPERA HOUSE—THE EVALUATION OF OBJECTIVE TESTS OF 'ACOUSTICS' OF MODELS AND HALLS
   Vilhelm Lassen Jordan 55

7. ACOUSTIC MODELLING—DESIGN TOOL OR RESEARCH PROJECT?
   Alexander N. Burd 73

8. ACOUSTIC SCALE MODELLING MATERIALS
   B. Day 87

9. THE REVERBERATION PROCESS AS MARKOFF CHAIN—THEORY AND INITIAL MODEL EXPERIMENTS
   R. Gerlach 101

10. THE INFLUENCE UPON AUDITORIUM REVERBERATION TIME CAUSED BY DIFFERENT POSITIONING OF ABSORPTION WITHIN THE FLY TOWER
    Paul Newman 115

Specific Types of Auditoria

11. VERN OLIVER KNUDSEN—A MEMORIAL TRIBUTE
    R. W. B. Stephens ... 119

12. THE ACOUSTIC DESIGN OF MULTI-PURPOSE HALLS
    Heinrich Kuttruff ... 129

13. DIFFERENT DISTRIBUTIONS OF THE AUDIENCE
    Lothar Cremer ... 145

14. THE FUTURE OF THE AMPHITHEATRE
    Sandy Brown ... 161

Contemporary Research

15. ASSISTED RESONANCE
    Peter H. Parkin ... 169

16. PROBLEMS OF SOUND REFLECTION IN ROOMS
    Zyun-iti Maekawa ... 181

17. NEW RESULTS AND IDEAS FOR ARCHITECTURAL ACOUSTICS
    M. R. Schroeder ... 197

18. THE CORRELATION BETWEEN SUBJECTIVE AND OBJECTIVE DATA OF CONCERT HALLS
    H. Wilkens and G. Plenge ... 213

INDEX ... 227

1

OPENING ADDRESS

V. B. TORRANCE

*Department of Building,
Heriot-Watt University,
Edinburgh, Scotland.*

We have here, with us today, representatives from some 22 different countries and among the participants and speakers are many of the foremost, most eminent specialists in this subject. I might add here that all of the specialist speakers have come to give of their expertise at this Symposium by specific invitation and I, together with my colleagues in the Department of Building, am honoured and indebted to them for coming here and giving of their time to making this what promises to be an outstanding gathering. From what we have seen of the subject material which will be covered by the various contributors, it is certain that the contributions will be of the very highest quality, as judged by any standards.

The last ten years have been a significantly productive era in architectural acoustics in which major advances have been made and in which important new developments have taken place. In the next two days, we shall hear many details of these. Indeed, Professor Bolt, in his address, will highlight a number of these important areas.

This Symposium, which incidentally is also notable for being the first major international symposium to be held by the University on this its newly established and developing campus, has been made possible by a series of events which I feel I should acknowledge.

Although the organisation of this event began just a little too late to be listed as one of the formal, official satellites of the 8th International Congress on Acoustics which has just concluded, we are indeed an important satellite of that Congress. A fact which should be acknowledged here is that although this gathering is, to the best of our knowledge, the first International Symposium on Architectural Acoustics to be held in recent times in Britain, the presence here of so many eminent specialist speakers would not have been fully possible without the holding of the Acoustical Congress.

However, one other reason of some importance for us in holding such a symposium has been the involvement by my Department in a modest, but what we hope will be a significant, manner in the design of the Edinburgh Castle Terrace Theatre complex. We are very glad to have among our contributors and participants, several persons connected with this project including Sir Robert Matthew, Mr. Sandy Brown, Mr. Tom Sommerville, Mr. Brian Annabelle and Mr. Stuart Harris. In addition, we are also very glad to have with us other leading designers associated with the planning and design of other major auditoria in Britain.

It was with deep regret and a profound sense of grief that we learned of the tragic and untimely death of Professor Vern Knudsen. Professor Knudsen was to have delivered a paper in person at this gathering and he had written to us saying how much he was looking forward to coming here. However, it is not for me to elaborate further here since Dr. Raymond Stephens, the President of the British Institute of Acoustics, has very kindly agreed to deliver a memorial address in tribute to Professor Knudsen during the course of our proceedings and we are very grateful to him for agreeing to do so.

2

Acoustics in Perspective

Sir Robert Matthew

*Robert Matthew, Johnson-Marshall & Partners,
Edinburgh, Scotland.*

May I first of all congratulate the University, and particularly Professor Torrance, on their initiative in promoting this conference. The title 'Architectural Acoustics' brings together, on the one hand, a field of study that in my own life-time has burgeoned, as Dr. Bolt will shortly tell you, from a somewhat tenuous connection with the past, into a highly sophisticated applied science, and, on the other, one of the oldest of man's activities, namely Architecture, the art of building.

Recently, thinking of this occasion, I enquired of the RIBA Library what major seminars or discussions had been promoted by that body on this subject in the last few years. The answer was that no such occasions had taken place. This is therefore all the more welcome and, both as a practitioner and as an ex-professor of Architecture, I am glad to have the opportunity to meet so many well-known experts in the field. I am here, of course, not as an expert—I suppose the specialists would perhaps call architects 'generalists'.

In addition, as you will hear later, the University is at this moment helping the design team, of which I am part, for the new Edinburgh Opera House, and Sandy Brown, Tom Sommerville and their colleagues in particular, by building and testing models at various scales of the main auditorium.

I interpret the title I have been given as a broad introduction, not to the scientific aspects of the study and application of the science of acoustics, but to a particular context—in this case an architectural one. I will be followed by an expert review of the subject; I hope I will not trespass, but this is extremely unlikely, knowing Dr. Dick Bolt and his profound knowledge.

Dr. Bolt gave us the benefit of his great knowledge and unusual range of talent in the days when the L.C.C. decided to have a new Concert Hall on the South Bank, and he first gave me a glimpse of the wide developments in the study of acoustics related to many activities, some quite remote from buildings, with which he was concerned at Southampton. I am particularly glad to see him here today.

I suppose that the kind of perspective, or context, I should sketch out is the general field of architectural design in so far as acoustic considerations apply. This of course covers many kinds of architectural programmes, as few building types do not have sound problems of one kind or another.

I remember, many years ago, just after the War, attending on behalf of the RIBA, a seminar in Turin, I think sponsored by UNESCO, on the design of auditoria. It took place in a building located in a narrow street, through which ran heavy traffic, accompanied by the usual exuberant Italian street dialogues. By the end of the first session, the foreign delegates, some of whom had come some distance to attend, had risen in a body to protest at what seemed to them an intolerable external noise level—a consideration that the promoters had evidently entirely failed to anticipate—and one, indeed, that the Italians who seemed to be accustomed to much more noise, found rather hard to understand.

Today, I am glad to say, we can hear each other speak without undue strain, in an environment sympathetic to the functions it encloses, that is, in so far as the relatively mean financial resources allocated to British Universities for their buildings will allow.

This brings me to the first contextual consideration—a very obvious one—buildings have to perform simultaneously in many different ways. Decade by decade, standards of expectation by users seem to rise. No doubt, these expectations may be, from time to time, encouraged for reasons not wholly related to the real well-being of people, as for example in fixing standards of artificial illumination, but, apart from that, standards of convenience and safety constantly strain financial resources that, far from keeping pace with these expanding requirements, seem to fall short at an accelerating pace.

All this puts the architect in some difficulty. He has, somehow, to balance the budget, and perform a kind of juggling act between a number of factors—space, stability, convenience, comfort, safety, for instance. In the last few years, in this country, and probably in others, building standards have been falling appreciably. It has become more and more difficult to meet exacting briefs and priorities are not always easy to establish. In auditorium design, for instance, acoustics' requirements have to be balanced with the requirements of theatre consultants concerned with sight-lines, 'theatrical character' and so on.

Priorities, of course, fall into more than one category. Some are fixed by the brief and the type of the building, sometimes also by the nature of the site, and this may well earmark considerable parts of the budget in advance, before the design process even starts. A simple illustration of this I have met on more than one occasion is a building located in an area—to quote our town-planning legislation—of 'special architectural or historic interest', where the town planning authority lays down as a condition of approval that the exterior of the building must be faced with, let us say, stone or other relatively expensive building material, without, of course, that authority offering to contribute to the additional cost involved. This may leave the building owner, especially if it is a public body like a University or a Hospital, with less resources than it had expected to design and build the interior.

These hazards and constraints we are well accustomed to in this country —hence the development over the last 25 years of the new art—I say art, but some might say it is a dodgy kind of art—of cost analysis and cost control, under which, at all stages of design, a balance sheet, so to speak, is drawn up and continuously reviewed, of all elements in the design—and what goes into one element may well have to come out of others.

Sound control is an element of this kind, part of the total environmental bill. This seminar will be mainly concerned with the special problems of auditoria, but adequate sound insulation, in its many aspects, may well be both critical and pricey. In some programmes of public building in this country in recent years, insulation has, I am afraid, been one of the casualties in the battle against inflated costs, usually much to the discomfort and irritation of the subsequent building users.

So the first point I would make in the general context of the subject is related to the fixing of priorities. While some of these are fixed and established before the architect comes on the scene, there are others on which he will be expected to give judgement, and that brings me to my second point. As buildings get more and more complicated, design becomes more and more a team exercise, very different from the practice of a generation or two ago, when team work was little known, and the more prestigeous the building, the more the individual architect considered it his right and duty literally to draw almost every line and certainly to design every detail.

Today, while the architect is still regarded as having prime responsibility for designing the building, and seeing it is built to his instructions, he has the additional task of integrating the inputs of all members of the team to achieve a result that, to his mind, meets substantially both the letter and the spirit of the brief. He has also the further responsibility, not always explicit, but I believe now written into the professional code of the American Institute of Architects, of safeguarding—indeed enhancing—the quality of the public environment, if I can use a clumsy phrase to describe an intangible, but nowadays politically a very real consideration.

In strictly legal terms, I suppose a design team as such does not exist. Contracts can be taken between client and architect only, with others then selected and appointed by the architect. They can equally well be taken by the client with a string of specialists, as in the case of the Edinburgh Opera House where, by the accident of history, the architect was appointed last. Functionally, the team may have some formal shape and meet regularly, taking minutes and so on. My own experience has ranged from one extreme to the other.

The design of the Royal Festival Hall in London 25 years ago was, surprisingly, an illustration of a large official organisation working with the minimum of formal procedures. There were many consultants, but no very systematic pattern of meetings. This was partly due to the fact that the L.C.C. had never built a building of that kind, but even more to the short time available to design and build. The brief had to be built up as one went along, and the various tests and experiments made in a number of directions had to be slotted in as best they could, partly during the building process itself.

Two very interesting and contrasting situations arose, both in relation to the design of the main auditorium. What was the auditorium for? One of the few directives given by the Council was that it was to be a Music Hall and nothing else. That was fine (of course, once it was built, it was used for all kinds of other things, some of them obviously not very successfully) but then the question arose—for what kind of music? Two eminent musicians were consulted—Sir Ralph Vaughan Williams and Mr. Benjamin Britten—plus Mr. Ralph Downes of the Brompton Oratory, to specify the organ. To cut a long story short, eventually Vaughan Williams resigned, as there seemed no way to reconcile differing musical viewpoints, and, incidentally, one very eminent firm of organ-builders considered Ralph Downes' specification unacceptable and refused to tender.

Here, I suppose, given time and an unlimited budget, it might have been possible to find a solution to meet all viewpoints—but, as neither were available, decisions had to be made, as best we could, and one view prevailed.

The other contrast, of an entirely different kind, arose within your own field. I well remember the occasion on which both the Building Research Station Acoustics Team and Mr. Hope Bagenal arrived for the first time together on the South Bank. Bill Allan and Peter Parkin arrived with something that looked like a caravan—on inspection it was full of apparatus, mysterious and complicated to a mere architect—valves, switches and wires, set up to measure, in a block house erected for the purpose, the noise of the trains going overhead on Charing Cross Bridge and, underground, in the tubes that ran under the site. This was splendid, functional and scientific.

Then Mr. Bagenal appeared with his equipment—in total one cardboard tube of the kind that tracing paper is rolled round, and, I think, a small tuning fork, but with a very sharp look in his eye and a vast amount of experience in his pocket. These two extremes came together and worked throughout in what I believe was a fruitful and helpful collaboration. As Mr. Parkin fired his pistols at an expectant but slightly awed audience, and Hope Bagenal brought along a friend who piped from the platform on a pipe not much larger than a fountain pen, we all, I think, realised the unique value of that joint team, unlikely ever to be repeated.

I read with some interest Dr. Bolt's summary of his review and his three phases. I suppose the Festival Hall would fall, perhaps a little uneasily, between the first and second—a preoccupation with reverberation, but also the beginning of the new concepts and techniques that came into widespread application in the last decade or so.

Today, we have a range of knowledge and experience to call on that 25 years ago was only just discernible; in our recent experience of designing the main auditorium for Edinburgh, we have Sandy Brown and his colleagues, with Tom Sommerville, who had already worked with the City Architect and his Deputy in the formulation of the brief. Very quickly we got from the Acoustics Consultants clear parameters for the geometry of the auditorium. These arose as a result of an international seminar organised by Edinburgh Corporation, in 1972, directed towards the problems of the proposed Opera House, and attended, I believe, by some of the speakers here today. These parameters then had to be quickly absorbed and understood and then combined with another set of requirements formulated by the theatre consultants, so that a preliminary design could be sketched out in sufficient detail to allow the testing models to be built.

These are now in the University; the 1/50 scale model has been constructed and spark-tested, and I am glad to say has been pronounced generally satisfactory. If it had not been, we would have been in some trouble. We are now working on the details and at the same time a 1/8 full-size model will shortly be built for further testing.

In the case of the Edinburgh design team, we have regular and formal meetings, with all the Consultants—Theatre, acoustics, structural and service engineers, quantity surveyors, and the client, represented by Mr. Stuart Harris, Deputy City Architect, who is also the Project Manager. We have only been together as a team since last September, but we seem to have come together and settled down well on a working basis.

The obvious advantage of a certain amount of formal structure for team-working is that all consultants are kept aware of developing thought in every sphere, and possible conflicts of interest are laid on the table and thrashed out before crisis situations arise. Good communications are essential, involving a fair amount of exchange of paper, but this is well worth the trouble.

While I might categorise team-working as a constraint, it is a constructive constraint, and can demonstrate what Buckminster Fuller would call the principle of 'Synergy'—a total result and a new value rising from a combination of separate elements whose constituents taken separately would not readily predict such a result.

It is, I believe, one of the essential roles of the architect to promote this synergetic process, but, if the whole turns out no more than the sum of its constituent parts—in terms of building—it is likely to be no more than the proverbial committee-designed horse-camel!

The fact that we have only too many architectural camels in the cities of the 20th century may well be, in part at least, a reflection of an uneasy transition from the days of the architectural prima donna who saw the design of a building only through his own eye, to the conception of an integrated and sympathetic team inspired towards a common goal.

The two qualities I have just indicated—sympathy and inspiration—probably lie at the heart of the matter. The sometimes astonishing quality of some alloys do not appear at all, unless in a chemical and physical way, the constituents are highly 'sympathetic'.

How much more, in the human situation, is this true. As for inspiration—this, surely, means more than the quality of leadership, although this in itself is indispensable. For too long we, and this certainly includes architects, have been persuaded to regard functional performance in 'low-order' terms, and to leave the rest—Sir Henry Wootton's quality of 'delight'—so often quoted, so seldom achieved—to look after itself.

The stringent cost-limits we are so often bound to have taken no heed of quality, so we can hardly be surprised at the sometimes violent reactions by the public against contemporary buildings.

You are all concerned with quality—its measurement and achievement.

You have the means, not only of measuring many things until recently thought incapable of measurement, but of understanding and observing qualities, perhaps only little recognised, for instance in the field of music, even by composers and musicians themselves.

We, as architects, are glad to use your insights to the maximum. If, at times, we appear unwilling or hesitant to go all the way, well, we are either lacking in grace, that is, understanding, or else, and I hope this is more typical, we are so beset by conflicting environmental demands that compromise of some kind becomes inevitable.

We are all concerned with quality, and the satisfaction of human need, and that we must understand. There is one advantage in growing old—one has the direct experience of change, and we have had more than a fair share of change in the last half century. The urban world seems to be getting noisier but the capacity to take noise, especially on the part of younger people, seems now to be enormous, and quietness, in some circumstances, may even be an irritant.

The new Library at Edinburgh University was to me an environment highly appropriate to its purpose—and quietness had obviously been high on the architect's list of priorities. Shortly after it was opened, the Librarian told me he had some complaints about the level of quietness, mainly from foreign students, and a demand for some continuous background noise—perhaps an indication of changing habit, but to me, and this of course well and truly dates me, it is a strange kind of reaction to a place of study and scholarship.

The dilemma we are constantly in, especially in designing buildings for public use, is that people can be vastly different in their reactions, and we have to work mainly on averages.

As an old teacher of architecture, I acknowledge readily that for some reason that might well be worth investigating, architectural students have, in the past, tended to construct round themselves a special kind of cocoon, within which they develop their own idea as to how people live and react to their environment. There was a well-known tendency to assume that most people are, by age, physical condition and temperament, during the whole of their lives, perpetually young and fit, and, should, therefore, in the way buildings are used, behave accordingly. This kind of outlook has, to some extent, been carried on into practice—sometimes with some quite disastrous results. I have put this phenomenon optimistically in terms of the past—in the last decade or so, I believe that the cocoon has worn thin, and, by one means or another, architectural courses, while bringing some elementary understanding of basic physics into environmental planning, have, at the same time, shed many of the traditional constraints to the understanding of human behaviour.

I once heard Julian Huxley on the radio say, 'The only generalisation I can make about the human race is that people are infinitely various'. The variety of human reaction is, to me, constantly astonishing, particularly of course in our own line, in the reaction to buildings.

And, if I can go back to the Festival Hall to conclude this introduction, one of the test concerts was designed to coincide with a meeting of a number of eminent consultants from several countries. We arranged their seating in various parts of the auditorium, and I sat well at the back.

Two tube railways to Waterloo ran directly under the Hall, at a depth of about 80 feet. A distinguished foreign acoustic expert had been in the hall shortly before the test and had mentioned to me the sound of the tubes, so I arranged that every time he heard something he would put up his hand. This he did, in the course of the test lasting about an hour, 23 times.

At the discussion session held after the test, I then asked the assembled acoustics men how many had heard the underground trains—only one then put up his hand. Incidentally, and I do not know if this had anything to do with the case, I noticed that he had exceptionally large ears!

Perhaps, nowadays, your sophisticated, and finely-tuned apparatus can deal with some, at any rate, of the range of human variety—I do not know.

Sitting in the theatre at Delphi, and hearing the strong and much delayed echo off the adjoining rock face—an echo that virtually gives two performances for the price of one—one acknowledges that objectives, in accordance with the universal desire of people to be themselves, can at times be very different.

And here, of course, we are in the realms of the formulation of the brief; this can be done in various ways. We would like to think that in this country, at least, the public authorities have advanced a longish way, first in examining what is required, and, secondly, expressing it in a way that can be understood by the design team. I believe the view is now widely accepted that the successful writing of the brief is in proportion to the extent to which the design team itself takes part in the process. This is not always easy—above all it depends on maintaining a common language between layman and expert—indeed between expert and expert, or, if I may go back to my original definition, between expert and generalist. Science can be a great and indispensable aid, but, as architects, we must be able to convey, in simple terms, what we are getting at—so please allow for the limitations of the generalist, who has his own problems of presentation.

I, and I am sure, all architects, must be grateful that science, in its many applications, is coming more and more powerfully to our aid; we are anxious to take full advantage of the new knowledge you can contribute. We greatly appreciate occasions like this, to meet, to discuss and probably to argue, above all, to understand one another.

3

Architectural Acoustics—a Review of this Century

RICHARD H. BOLT

Bolt, Beranek and Newman Inc. Cambridge, Massachusetts, U.S.A.

Now it was by deliberation that I chose to refer to the first quarter century and the second quarter and the third, because of the very well known principle, the uncertainty principle, which is that the narrower the gap of time about which you choose to speak, the more detail you must give and the more you must know about what you are saying. So let me very quickly just refer to the four quarters of this century and then maybe make a couple of comments which are very much in line with what Sir Robert has said. What I wanted to do is just to under-score the essential sentiments that he has already expressed so very well.

That first quarter of the twentieth century started off with something like a bang, although I am not sure that Sabine ever did use pistols for checking reverberation. He went around with his organ-pipe and by the time that quarter century had ended, roughly 1925, some interesting things had happened, and one was that we knew all about acoustics. We had a reverberation equation and it was 0·049 V/A if you believe in English Units. There was not yet that terrible complication brought in by Professor Eyring, although Professor Eyring was already in the field of acoustics and was teaching people with names like Knudsen and Fletcher. So I would characterise this first quarter century as being largely one of reverberation, since just a few years before 1900 Sabine put together his reverberation equation. Before that, although human beings had had some creative ideas about sound and echoes and intelligibility, no-one had really put together the relationships that say if you can measure quantitatively how long sound lasts you can make a direct correlation with how well you like what you hear if you are listening to music or speech and how well you like the acoustic environment with which you are listening.

But it would be unfair to Sabine to suggest that all that he did in that first quarter century was to contribute the reverberation theory. As may be seen from reading Sabine's own writings, he was extraordinarily perceptive about most of the aspects of acoustics qualities that we look at today. He produced some very interesting pictures showing standing waves of steady state sound, of more or less one frequency, in a room, and he recognised that the complexities of standing waves in rooms, especially in smaller rooms, would have to be dealt with somehow. He certainly understood the importance of criteria to relate human judgements, responses if you will, to the physical stimuli, to the objective acoustical qualities. He ran what I believe was the first series of tests reported anywhere in the literature, in which he got several friends of his to go into several different rooms in a building and listen. Meanwhile he was measuring reverberation time with his organ pipe and he rank-ordered these rooms as they were liked; 'How well did you like this room and this room and this?' By undertaking such an experiment he ran, I think, the first psycho-physical judgement test and came up with some ideas about optimal reverberation.

Of course he was also interested in noise intrusion, though he did not have a long series of specialised terms such as N.C. and T.L. and so on. Today we have a list of 10 or more different kinds of curves and scales with which we measure the noise for different conditions. But he certainly was sensitive to noise, as anyone who has been in the Boston Symphony Hall will have recognised. There the inner hall is surrounded completely by stage areas, by entry foyer, by side halls. The whole thing clearly was an effective job, considering it was designed 75 years ago. Although this aspect of the design was originally intended to exclude the noises of horses' hooves on the cobbled stones outside, it is interesting to note that even the noise of the occasional aircraft flying overhead today does not cause too much trouble inside of that hall.

To conclude about the first 25 years, Sabine did, I think, a remarkable job of laying the ground work of architectural acoustics as we understand it today, but I would be inclined to say that the first quarter century ended with the subject being very simple. We understood exactly what to do. We put acoustic tile everywhere, especially in all the lecture rooms. We put acoustic plaster in the churches.

Now, at just about the turn from the first to the second quarter, specifically 1923, 4 and 5, I was a boy soprano in a church in the States, and during that period the building was changed from a hall which was just lovely to sing in over to one of these brand spanking new acoustically treated halls with tile all over it—and suddenly my voice was lost. So we made our mistakes too, but I would say that the first quarter century was characterised by us ending up knowing all about acoustics.

Now I am going to try to convince you that the second quarter ended up with everything being the opposite extreme, extraordinarily complicated, too complicated, and at the end we felt we knew nothing about acoustics. Just think of the things that happened during that second quarter. Prior to that, about the only people involved in acoustics were, of course, Hope Bagenal with his little tube, Wallace Sabine and his relative, Paul Sabine, Professor Watson at Illinois, and a few others. Then in the 20's, in fact just around 1925–7, young Knudsen went to Bell Labs. and later went to Hollywood, and he showed up there just about the time that sound was showing up in movies. Professor Knudsen very quickly gained experience in some of the complexities and difficulties of acoustics of studios through having designed and supervised the construction of the first sound stage and studio in connection with the movie industry in Hollywood.

Well now, my little bit of history here is going to be somewhat vague and very incomplete. I should be naming 10 names for every one I do, but I just want to touch the highlights and point out that early in this second quarter came quantum mechanics and Morse into acoustics, and sound waves in rooms, and some quantification of the standing waves that Sabine had measured in rooms.

Then we went through a period from middle-thirties to late-forties getting deeply into normal modes of vibration, normal frequencies, statistics, standing waves of all kinds, influence of room shape, acoustic impedance and not just absorption coefficient, complex acoustic impedance, how it varied with angle of incidence, how it varied with the structure of the graininess of the material, and even how that which you measured in the way of absorption coefficient was not really constant because under some definitions the value could depend on where you put the material in the room and how large the material was.

I think I only need to say that when 1950 arrived we had turned architectural acoustics into an extraordinarily complicated subject. We had dug very, very deeply into many of the details of the physical behaviour of sound waves in rooms of all kinds. Also during that period we had made our first really deep investigations into the psycho-physics of the problem. We had come up with, for example, speech intelligibility tests and various other studies of interactions between humans and their sound environment. By 1950 we had very deep knowledge and understanding about most of the complexities of the subject of room acoustics. However, we hardly knew how to apply what we knew. Although we were doing our best, we were struggling and there were not very many people applying this knowledge.

I would characterise the third quarter century, the one we are just now closing, as being one in which we have regained some balance. We are ending this third quarter without either an over-confidence in the simplicity and the completeness of our knowledge, as we did in 1925, or conversely, in desperation of the subject being so complex that we could hardly ever apply it. In the past 25 years we have seen applications of these principles to a very, very large number of buildings. As a result we have gained some real confidence. One of the principal reasons for this confidence is a technique that we have now superimposed upon our more theoretical and analytical studies, the technique of acoustic modelling. I have especially looked forward to being at this symposium in order to hear the up-to-date discussions about models and what we can do with them.

As long as I am referring in very broad terms to quarter centuries at a time, I can of course exercise my privilege to say what the year 2000 will look like. First, however, we need to look at some historical data, so I have attempted to gather some data about the growth of enterprises such as acoustics and architecture. I have been unsuccessful in getting a coherent set of data to describe the entire world, but I have some reasonable data describing the United States. Our population has been increasing at 1·8% per year—but I find it more useful to give you the percent increase per quarter century, just to be consistent. Our population in the United States has increased this last quarter century by about 50%. Since 50% per quarter century is the average figure for the last three quarters, I should expect something like this to continue (perhaps modified downward).

Let us now look at the population of architects. It is not so easy to define because, although we have professional registration we also have a significant number of people who hold degrees in architecture who are not practising it in a formal sense, and we also have many people practising design of buildings who are not formally architects. However if we take a very simple approach and look just at the membership of the American Institute of Architects, we will find that it is increasing at 40% per quarter century. In fact a graph of the number of members of the American Institute of Architects is almost a straight line of 40% slope on a semi-log scale from 1900 to 1975.

How about the acousticians? Well, the acousticians are increasing at 270% per quarter century. This makes it perfectly clear that some day the acousticians will completely swamp out the architects! Of course we must look at more than just rates of growth. We must look also at absolute magnitudes, and here a slightly different story shows up. In the United

States today there are about 35 000 members of the American Institute of Architects and only about 5000 members of the Acoustical Society. So there are about seven architects to one member of the Acoustical Society. However, if we look at publications in the *Acoustical Society Journal* or other journals, or at papers given at the Eighth International Congress which has just ended, we will find that not all of them are papers on Architectural Acoustics. So the question comes, 'What fraction of acoustics is architectural?'. This is not an easy question to answer because, as we can see in the classified index of the *Acoustics Journal* in America, although one section is called Architectural Acoustics, there are other sections named Noise, or Instrumentation, or Psycho-acoustics, and other subjects of significant importance to the field of Architectural Acoustics. Well, I counted up a large sample of papers and gave each a weighting as to its architectural acoustics content. When I put together all these data I came up with a number that says about one-seventh of the activity in acoustics is Architectural Acoustics. Instead of there being a seventh as many architectural acousticians as there are architects, there are about a seventh of a seventh as many. So it is going to take quite a long time for the acousticians to swamp out the architects.

In this review I think we should talk quantitatively about the amount of acoustics that gets put into practice. What fraction of new buildings receives acoustic design? This is even more difficult than finding out how many people there are in the field or at what rate the amount of information is going up. However, by making various kinds of assumptions I have come up with a number. The new buildings include many residences, individual homes and multiple dwelling buildings, a number approximately equal to the number of non-residential buildings. Furthermore, the non-residential buildings include a large number of offices and commercial and industrial buildings, where no special acoustic design would be considered necessary. These observations lead me to estimate that only a very few, 2 or 3 per 10 000 buildings, receive explicit acoustic attention today, at least in the United States. However, when I attempt to project this number historically starting with a number in 1900 which is maybe just one building, I think by the year 2000, if present trends continue, the percentage of the buildings receiving some explicit attention in acoustics will have reached about 1%.

This one percent figure may turn out to be quantitatively inaccurate, but I think it is qualitatively correct. Only a very small fraction of all buildings will be receiving direct, explicit attention by professional experts in acoustics. Let me call this fraction the One Percent. Then what about the rest, the Ninety Nine Percent? These two fractions, the One and the Ninety

Nine, pose quite different problems for us as we enter the final quarter century, so I shall divide my remaining comments in this review into two corresponding parts.

For the One Percent, we should continue to apply our specialised knowledge in ever more effective ways. We will be working on problems that are very technical and detailed. We will be working on opera houses, concert halls, speech rooms, studios, and the many other kinds of spaces in which good hearing conditions are of significant importance. Already a substantial number of such buildings are receiving specialised acoustic attention in most countries. Already the architectural press gives considerable publicity to the acoustical component, and it would be an unusual architect who did not know about the existence of acoustics specialists.

So for this One Percent our problem will be to develop more effective ways to make our theories really prove out in practice. We shall have to work as close collaborators directly and intimately with the architects and owners and builders and suppliers of acoustical products. We shall have to work out more efficient and economic procedures for integrating acoustic design into the entire system.

We know much more about the ideal performance of a complex wall at 70 dB isolation in a laboratory than about how to achieve that number in practice, because of the long chain of events that can vitiate the value of that wall. We know much more about the technical aspects of the time variation and space variation of sound waves within the rooms than we actually apply. So here in this next quarter century our big problem is going to be to find out how to put into practice more effectively this tremendous knowledge that we have gained and upon which we have done a reasonably good job of codifying and simplifying.

The other major problem, the one that has to do with the 99% of the buildings that get no explicit acoustic attention, calls for another kind of professional responsibility. The medical profession takes upon itself a responsibility to help all human beings achieve and live with good health, where health is defined in broad terms. It seems to me we might as a profession of architectural acoustics specialists take on a responsibility for the acoustic health, if you will, of all the people in all our countries.

When the medical profession takes on this kind of responsibility, it does not attempt to have a fully trained M.D. physician go to every person every time that person breaks a toe-nail or has a cold or sneezes or gets a headache. Rather, what has happened through general education that has been encouraged by the health professions, is that most members of the public have some fundamental understanding of what health is, what helps you to stay healthy or what sorts of activities might be risky for you.

I think that we might give some serious attention in this coming quarter to helping all the people—and that is a very large order—acquire some basic understanding of the nature of sound, some basic understanding of the way in which it can make your hearing deteriorate, some understanding of the way in which good acoustical qualities in a home or in an office may contribute to happiness and a more satisfying life. Then I think that we will find that a very large percentage of the new buildings built are going to have some good acoustic principles put in, even those buildings which have not attracted the direct, explicit attention of a professional specialist in acoustics.

If we accept this concept, then perhaps we can redefine our roles as specialists, as acoustician-architects and architect-acousticians. We really have a two-fold role. One is that of working together to design, to the best of our ability, specialised facilities in which acoustic performance is of major importance. The other role is to distribute the knowledge, to simplify the basic information, and to help get that information into the schools and colleges and communication media. This should not be done in a specialised course in acoustics, but rather in a way which can best express and convey the basic ideas of sound and its interactions with people in such simple terms, that newspapers and magazines and television stations will want to put on programmes that will help in this process.

When I refer to us as technical specialists, as the 'surgeons' working on specialised problems, I am thinking of more than just the distribution of sound energy inside large halls. I am really thinking about the broadest definition of acoustics and the broadest definition of architecture, and about their intersection in which architecture is the art and science of creating spaces for human beings in which to live and work and function, whilst acoustics is the science and technology of mechanical vibratory energy in all media at all frequencies.

One of the newer examples of this intersection is the aero-acoustics of sky-scrapers, a very interesting subject indeed. You have wind blowing against a building at 50 miles per hour (80 kph) and perhaps going round a sharp corner and shedding eddies and creating turbulence and applying a random driving force to the structure. The building itself is a lightweight structure and in many of our sky-scrapers, to overstate slightly, the structures are becoming more like aeroplane fuselages than like buildings. With high 'Q' factors they respond much more susceptibly to driving forces, and in fact take on some of the characteristics of an aircraft. Well here is one type of application of basic acoustics knowledge to the field of construction.

Another area in which we acoustic specialists have not been as bold as we might have been is in helping to guide the architect and the builder in the selection of materials and equipment and facilities; in selecting the motors and blowers; in selecting the kitchen equipment that is going to go into the multiple dwellings; in selecting the furnace. If we were to help in this process, I think it would encourage more of our manufacturers and suppliers of equipment to improve the acoustic performance and reduce noise.

So I see the acoustic-architecture partnership in this next quarter century being involved in a much wider range of activities than it has been in the past—activities in which our specialised technical knowledge can apply. I see us together being involved in helping all the public to gain sufficient knowledge about acoustics so that the benefits of all of this study will become truely widely spread.

4

Design Considerations from the Viewpoint of the Professional Consultant*

PAUL S. VENEKLASEN

*Paul S. Veneklasen & Associates
Santa Monica, California, U.S.A.*

There are several questions of attitude and policy that must be faced and resolved by the professional consultant if he is to be effective. Whom is he really to serve? The architect who may choose and hire him? The owner who may have chosen him and surely pays the bill? Or, in the long range, the thousands of patrons of the arts served by the facility? Presuming that he should choose the latter for his real responsibility, he will immediately be in potential conflict with the first for his economic responsibility, and the second legally, while the third; *i.e.* the patrons, are the amorphous historical victims. Yet it seems to me that it is the latter who must dominate his concern.

Conflict between architect and acoustician seems to be a constant concern. Why is this? First there is usually a real difference in training and background: the architect is usually artistically, *i.e.* design oriented—a most essential influence for the project; the acoustician should have a scientific and engineering background. The acoustician is presumed to be myopically one-function oriented. The architect must coordinate all functions—which he presumes to be totally incompatible. The architect, at least in the U.S.A., rarely has prior auditorium experience; more rarely repeated progressive, evolving experience. The architect inwardly cherishes the chance to create a monument to his skill. The acoustician may be similarly compelled. The architect considers himself the ultimate decisive person, not only between all members of his team, but between his own ideas and those of each of his team. Each member of the team feels a dreadful re-

* This address is dedicated to Professor Vern O. Knudsen, eminent educator, administrator, builder, master acoustician, Christian gentleman, cherished friend, and the writer's principal mentor in this field.

sponsibility, hopefully to the project, but surely to his own reputation and future. These are all very compelling forces in human relations. Ultimate success comes only from great competence in each team member. Each should be highly conscious of these motivating factors and resolve to be highly respectful of the needs of each other. Compromise with really vital elements is very hazardous. In the event of unresolved differences or enforced compromise, the only recourse for responsibility is to go on written record. But I firmly believe that total compatibility is achievable. Success and satisfaction for the acoustician exist *only* when the architect is also well satisfied.

What image is the consultant to present? Is he the High Priest—giver of concise oracular advice?—I hope not. He should be able to present also a rational defense of his advice. We believe that the consultant should provide detailed design for all areas of his responsibility.

What is the certainty in acoustical design? Is it an Art—or a Science? I believe this is a specious question. An auditorium is a facility for presentation of Performing Arts. Therefore, the end product that we serve is Art. Our objectives and criteria must first come *from* the arts. Are these arts subject to analytical and quantitative formulation, prescription and criteria? I believe they are and this alone must be one challenge. It is a pre-condition that we have quantitative objectives if design is to be quantitative, *i.e.* scientific.

Suppose we are given quantitative objectives. Is acoustical *design* scientific? Can results be formulated, can they be predicted, designed, achieved and verified quantitatively? I believe all this is an essential challenge, and that we are slowly progressing. Perhaps the most difficult task for the scientist is to know and admit the state of progress: to fairly assess what is known; to be alertly aware of what may be unknown; and to admit the distinction. In short, acoustical design is aspiring and progressing as a science in the service of the Performing Arts. The remainder of this article illustrates these points.

Presuming, then, that our task is to furnish detailed design, let us outline and briefly explore some of the specific areas of responsibility. This must of necessity be only a brief survey of our design tasks.

First, the *programme* must be completely stated and mutually understood, all aspects being drawn out by conference and discrete inquiry with the owner. Figure 1 lists the subjects and problems. To list the kinds of performances; to avoid narrow utility; to avoid transient individual biases; to determine the size and range of sizes of audience; to impress upon the owner that audience capacity determines the attraction for touring groups

Performance Categories in a Multi-Use Auditorium

Lectures	Compatibility?
Travelogs	
Motion Picture	
Projection TV	Reverberation?
Speech Drama	
Music Drama	
Opera	Difficulty?
Recital	
Orchestra	
Oratorio	Achievability?
Pipe Organ	

Fig. 1.

and therefore the diet for the community. Two design essentials become immediately obvious: practical *convertibility* for varied performance types, and practical *contractibility* for various sizes of audience.

This points to another widely held conviction—that a multi-purpose design cannot serve any one performance type well. I am convinced this is a myth. In fact, I have come to the belief that the best concert hall can be achieved in the opera–stagehouse configuration.

Exterior noise must be excluded. The interior criterion is to eliminate

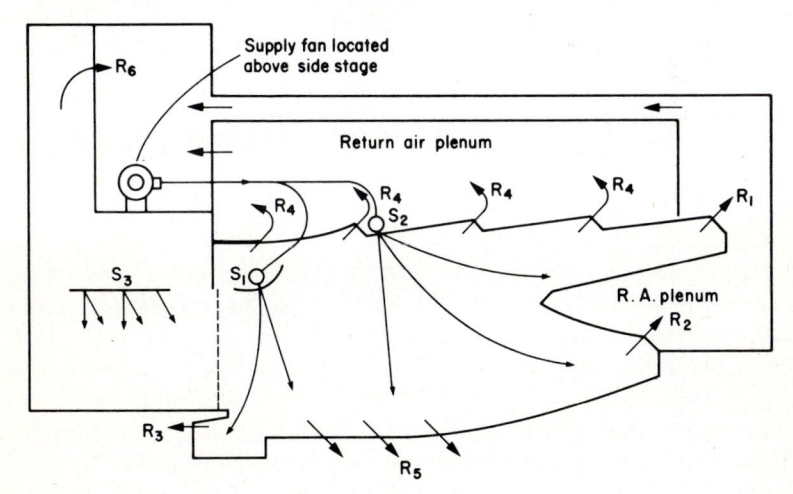

Fig. 2. Low velocity air distribution in an auditorium, for minimum noise and maximum comfort.

distraction for distinctive, recognisable sounds. Some of the sources are extremely difficult, and may occasionally intrude, such as overhead helicopters, or emergency sirens.

Mechanical noises from airborne and vibratory sources such as interior machinery, must be controlled. NC-15 criterion should be used. The most common failure is with ventilation noise—from ducts and aerodynamic turbulence. Very low velocity distribution is necessary. The air must be moved to the proper places by convection. Figure 2 shows our recommended methods.

Stage facility designs are very challenging, especially for the multi-use hall.

Fig. 3. Integration of demountable orchestra enclosure with canopy and general architecture to create concert hall form of multi-use theatre.

Figure 3 shows a demountable orchestra enclosure which converts the theatre into a recital hall. It should achieve: one-room continuity architecturally and acoustically; adjustable projection of sound to the hall; internal reflection for cohesive performance; adjustable projection or discard of sound to the stagehouse for balance control; flexible spacial enclosure for recitals, full orchestra, orchestra plus chorus; rapid convertibility—each piece operable by one or two men; self-contained lighting. Figure 4 shows the flexible area enclosure. Figure 5 shows the self contained lighting.

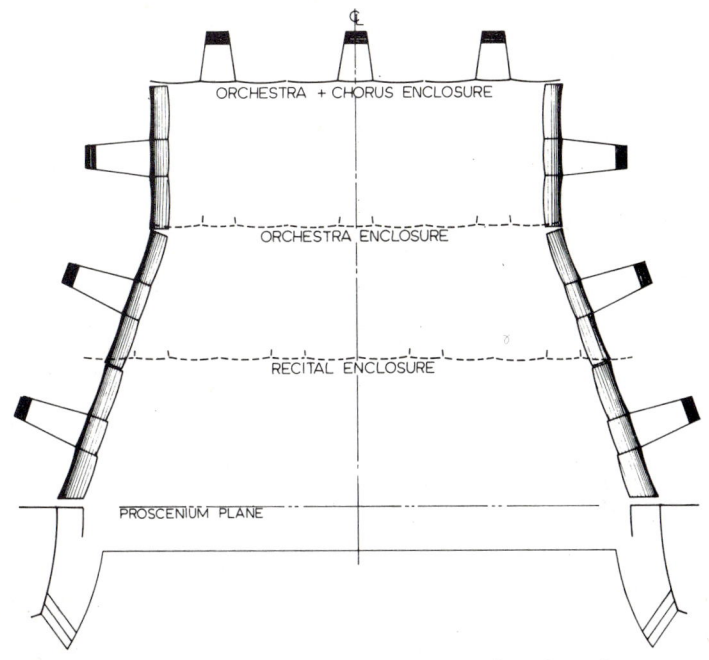

FIG. 4. Plan of flexible orchestra enclosure wall configurations.

FIG. 5. Self contained orchestra lighting in proscenium arch and ceiling panels of orchestra enclosure.

The orchestra pit is vital to convertibility and requires specific acoustical design. Our design has evolved to a degree in the same direction as the original Wagner pit at Bayreuth. It must achieve: musical balance between stage and pit performers; adjustable size; reflection of sound from pit to stage to pit for cohesive performance; physical and visual tenability for pit performers; musical balance within the pit orchestra. Recommended terracing is shown in Fig. 6.

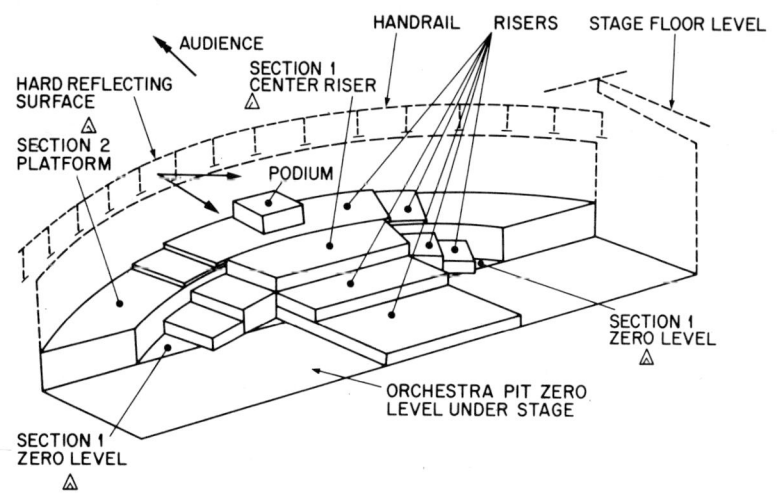

FIG. 6. Modern orchestra pit—partially submerged, viewed from under overhanging fore-stage.

A rather simple electro-mechanical device permits rapid operation of the smaller pit section, which in turn picks up the outer section as a fork lift.

The plan form and profile of the audience seating deserves analytical design. This may be approached through a quantitative analysis of effectiveness. Which form is best? the Greek arena? one or several balconies? the European opera form? We feel that this subject may be approached analytically. Sight lines and sound lines are interestingly compatible, as well as some psychological factors. We find five independent factors. These functions and their individual behaviour are shown in Fig. 7.

The row-to-row sight line clearance is the 'h' value. A high value favours easy viewing over heads. There is enormous sound absorption for sound progressing over audience at a shallow angle, so that a steeply sloping floor also favours sound propagation to rearward audience. Increasing distance from the stage is a negative factor, decreasing effectiveness. A seating position having a large eccentricity angle is less effective. Either too low or too high viewing angles are undesirable, and a steep balcony is also unsafe. Locations under a deep balcony overhang are undesirable, because the reverberant sound is lost, as well as the appreciation of the visual spaciousness. Each of these factors may be expressed quantitatively with an effectiveness function. For any seating location, the product of the five factors gives the effectiveness of the seat location. Average for any specified area or total area average may be calculated. This analytical technique permits comparisons of various plans or portions of seating plans. Interesting and somewhat startling comparisons result. For example, for rather large occupancies, the arena plan shown in Fig. 8 is more effective than a balcony plan (shown in Fig. 9) in spite of greater distances. A balcony plan is great for those in the first row of the balcony, but all other seats suffer for this advantage. Figure 10 shows that, for successive contractions, *i.e.* for smaller audiences in a large hall, the advantage is especially great. This kind of analysis needs further testing and development.

There are many important features to be achieved in acoustical design, the objectives are listed in Fig. 11. In contrast with the usual preoccupation with reverberation time, we would place it sixth in a list. Clarity is achieved by supplementing the sound travelling directly from the source with successive reflections from carefully designed ceiling panels spanning across the auditorium. These reflections preserve source localisation. Figure 12 shows that with these surfaces, the direct plus coplanar reflected sound can be kept quite uniform in level over the entire audience.

Freedom from strong discrete echoes and strong envelopmental sound can be achieved by designing a series of wall surfaces as shown in Fig. 13 to provide a series of controlled, successively delayed mini-echoes from many lateral directions. Figure 14 shows an echogram for a small theatre, indicating for several seat locations the series of echoes arriving from different directions with indicated delays after the direct sound. The plan of these walls must be adjusted to meet specific criteria at all seats. In constructing and assuring the results, we can use optical modelling and also acoustical modelling as shown in Fig. 15.

The *envelopmental* sound furnishes several important purposes. Some we understand, and others are still mysterious. A demonstrable physical feature is that the laterally travelling sound does not suffer the severe selective absorption due to the seat rows that occurs for the direct sound. Envelopmental sound displays the size and specific character of a hall. The purely emotional benefit of envelopment needs study. And the apparent

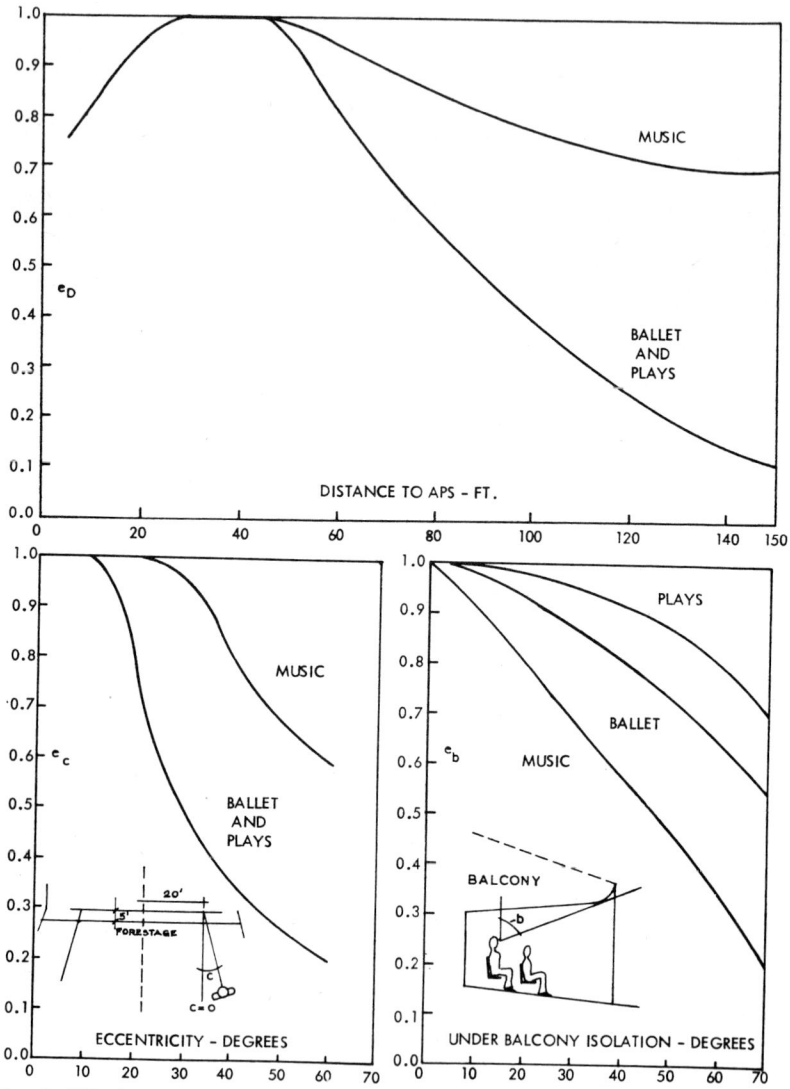

FIG. 7. Effectiveness factors for quantitative design and evaluation of theatre seating plans and profiles. (*See also facing page*.)

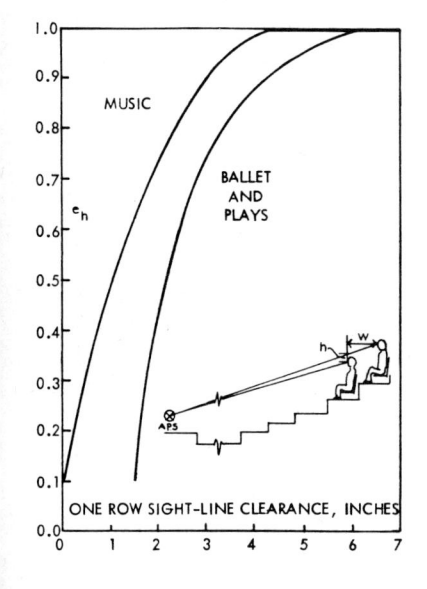

Design Considerations of the Professional Consultant

NOTATION AND EVALUATION:
SPECIFIC EFFECTIVENESS OF ANY SEATS:
$$e_S = e_D \cdot e_h \cdot e_a \cdot e_c \cdot e_b$$
AVERAGE EFFECTIVENESS FOR ANY ZONE:
$$E_Z = \frac{1}{Z} \sum_Z e_S$$

AVERAGE EFFECTIVENESS FOR AUDITORIUM:
$$E_N = \frac{1}{N} \sum_N e_S$$

Z = NUMBER OF SEATS IN ZONE
N = TOTAL NUMBER OF SEATS
$N = \sum Z$

FIG. 7. Continued

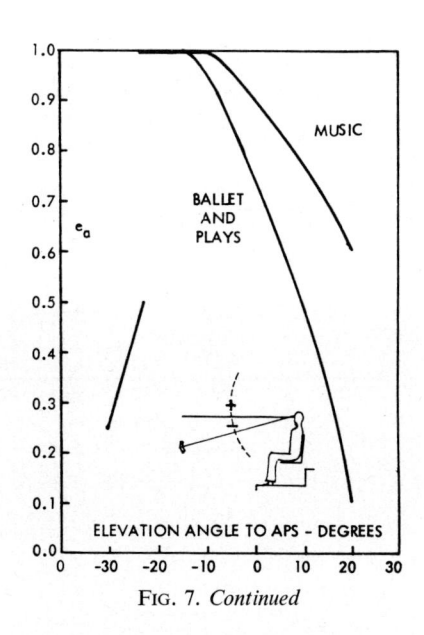

Half plan

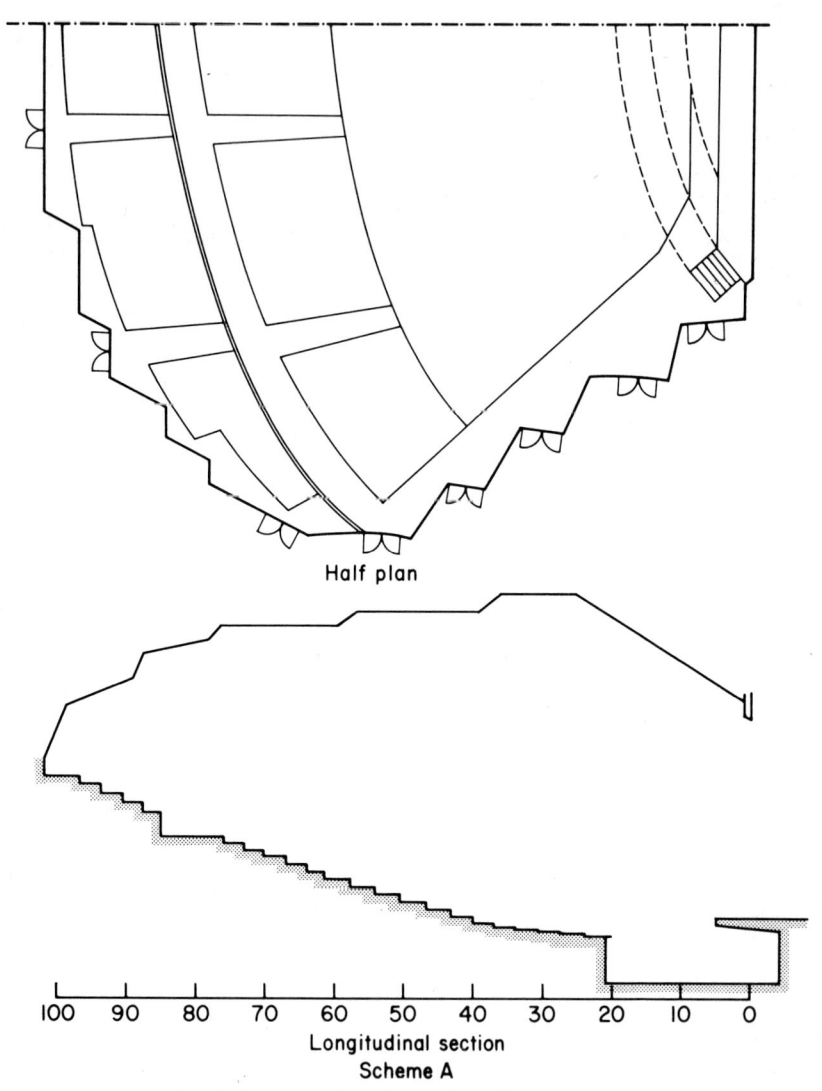

Longitudinal section
Scheme A

FIG. 8. Arena seating plan and profile—Scheme A.

Half plan

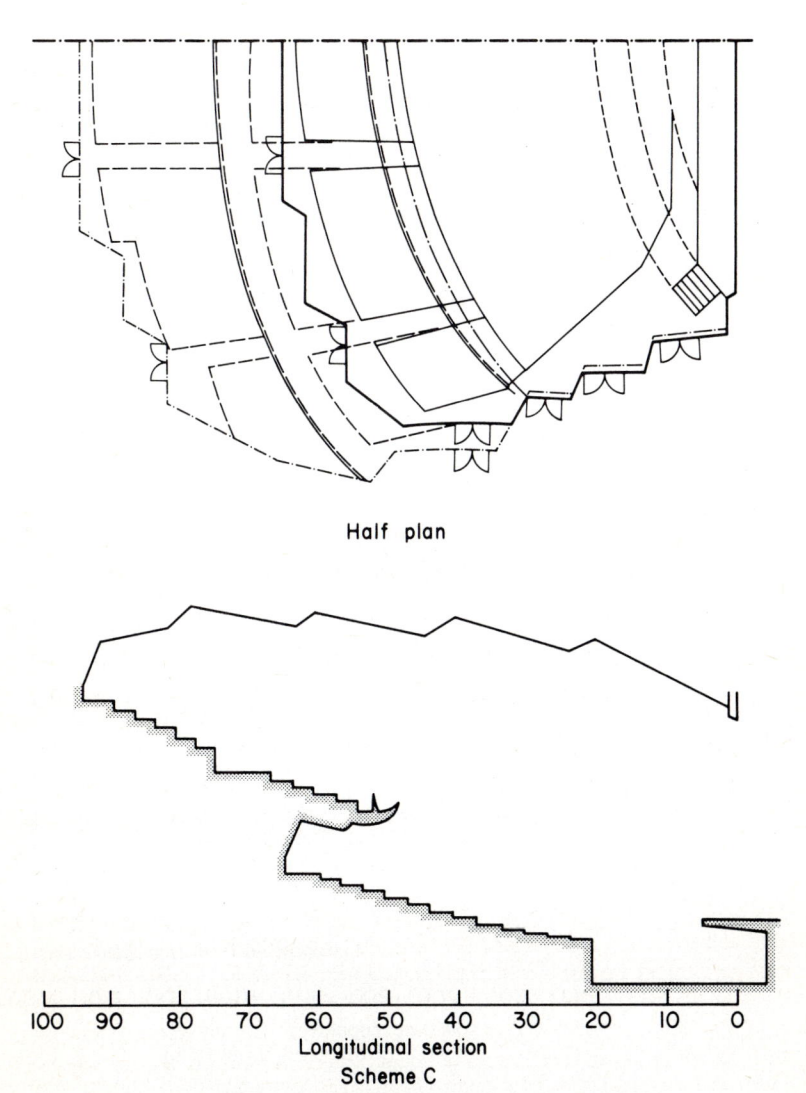

Longitudinal section
Scheme C

Fig. 9. Balcony seating plan and profile—Scheme C.

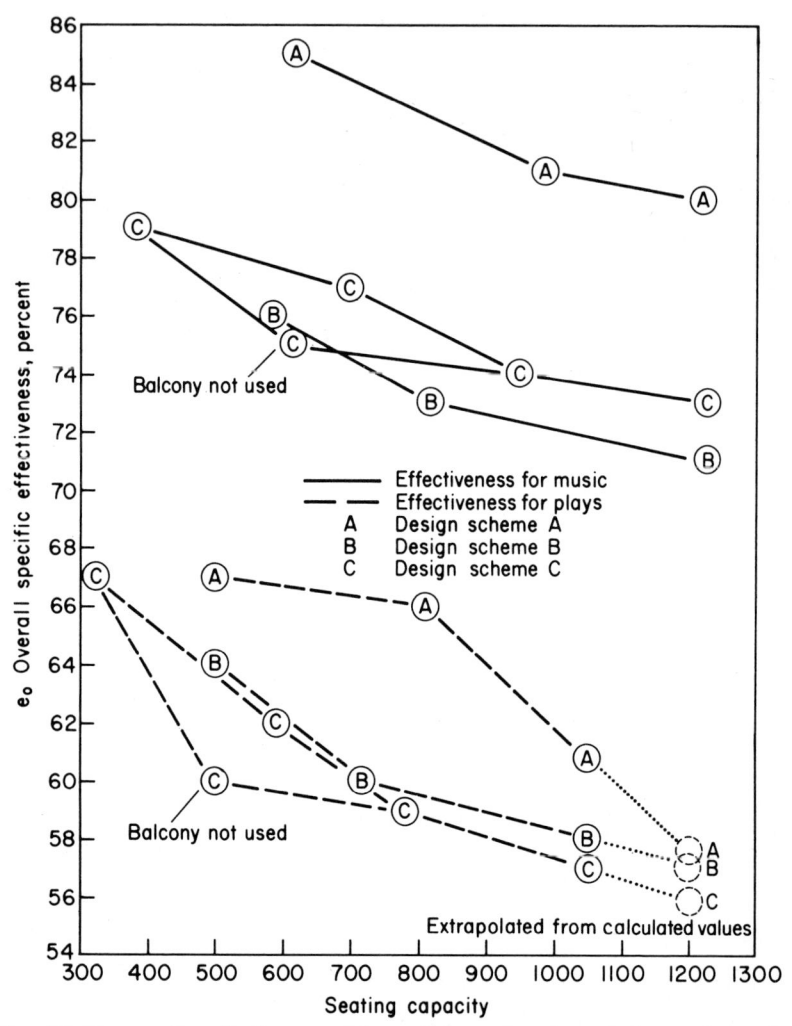

Fig. 10. Comparative effectiveness of Arena Scheme A and two balcony schemes for music and plays and for several degrees of audience contraction. (Auditorium—University of Washington.)

Acoustical Design Objectives

Clarity/Intelligibility	Strong, Uniform Direct Plus Coplanar Reflected Sound
Balanced Projection	Selective Control of Reflection From Stage to Audience
Cohesion for Performers on Stage	Control of Reflections on Stage
Freedom from Echoes	Control of Rear Wall Reflections
Strong Envelopmental Sound	Control of Side Wall Reflections
Reverberation Level and Decay Rate	Distribution of Interior Absorption

Fig. 11.

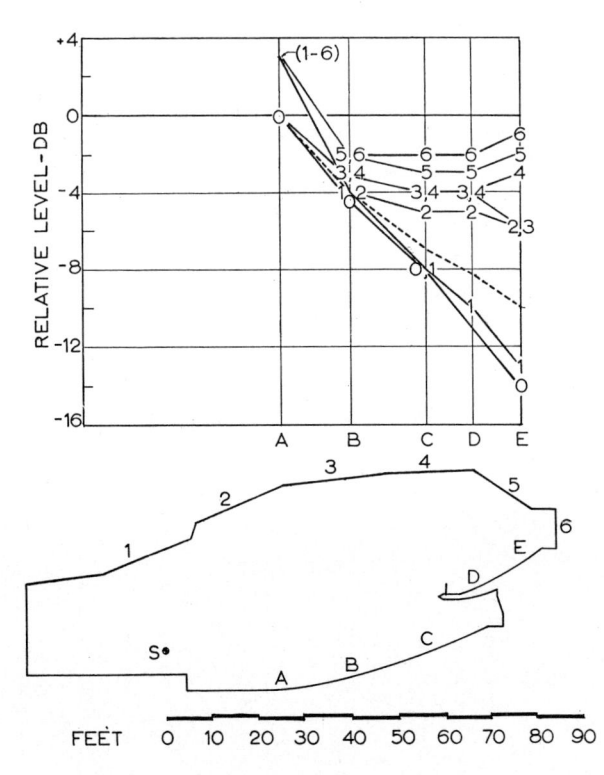

Fig. 12. The role of ceiling panel orientation in providing a near uniform level of direct plus coplanar reflected sound over an audience surface.

34 Auditorium Acoustics

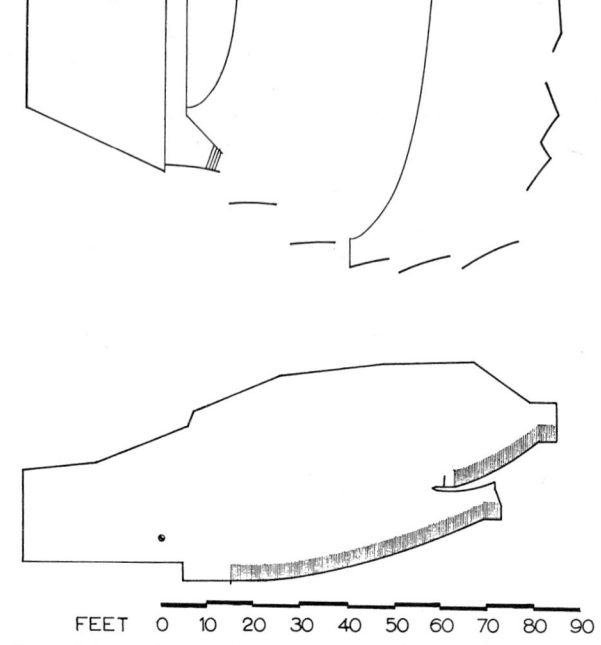

FIG. 13. Plan of side and rear wall panels to provide strong envelopmental sound.

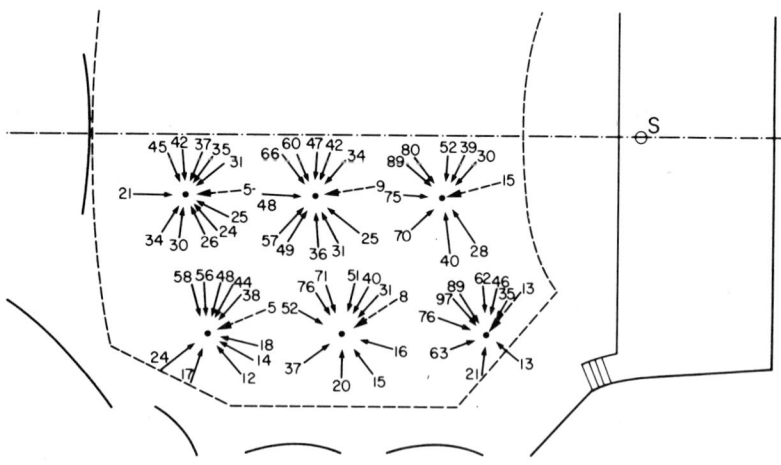

FIG. 14. Echograms at several audience positions showing directions and delays for envelopmental and canopy reflections relative to direct sound.

supplement to deep bass loudness from reflected repetition is a total mystery of hearing.

What should be the criteria for envelopmental sound? Here we have the results from a series of successful and cautiously evolved halls. The same acoustic echograms may be made in the finished hall as in the model. It is no surprise that, since the velocity of sound is the same, the correlation is quite exact. The full scale hall, of course, shows much more diffuse reflection; but the major large surface reflections show clearly.

FIG. 15. Use of acoustical modelling to confirm design of acoustical reflecting surfaces.

The reverberant sound is, of course, a compromise. The maximum tolerable reverberation time for speech and lyrics, even using electro-acoustic reinforcement favouring the direct sound, is about 1·5 sec. Orchestral music in the grandest scale is favoured by times of 2 to 2·5 sec, and much pipe organ music demands even larger times. Here is one way the stagehouse of the multi-use auditorium may be put to use. Depending on the amount of fixed stage drapery, the reverberation time of the stagehouse may be much larger than the hall. In a technique we call Stage-to-Hall Coupling we can transfer this superb, otherwise wasted, reverberant sound from the stagehouse, through two electro-acoustic channels from the stage into the hall, as shown in Figure 16. By this means the effective reverberation time can be varied by the turn of a knob over a range of 1·7 to 2·8 sec in our Portland Auditorium, as shown in Figure 17.

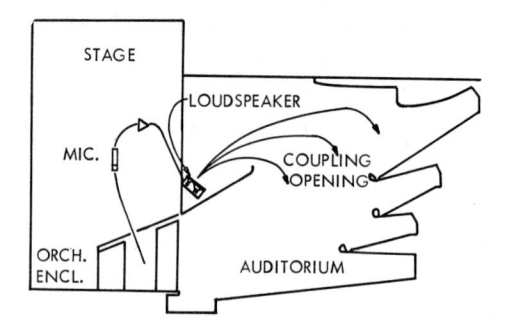

FIG. 16. Use of electro-acoustic coupling for stage-to-hall transfer of reverberant sound for adjustable reverberance.

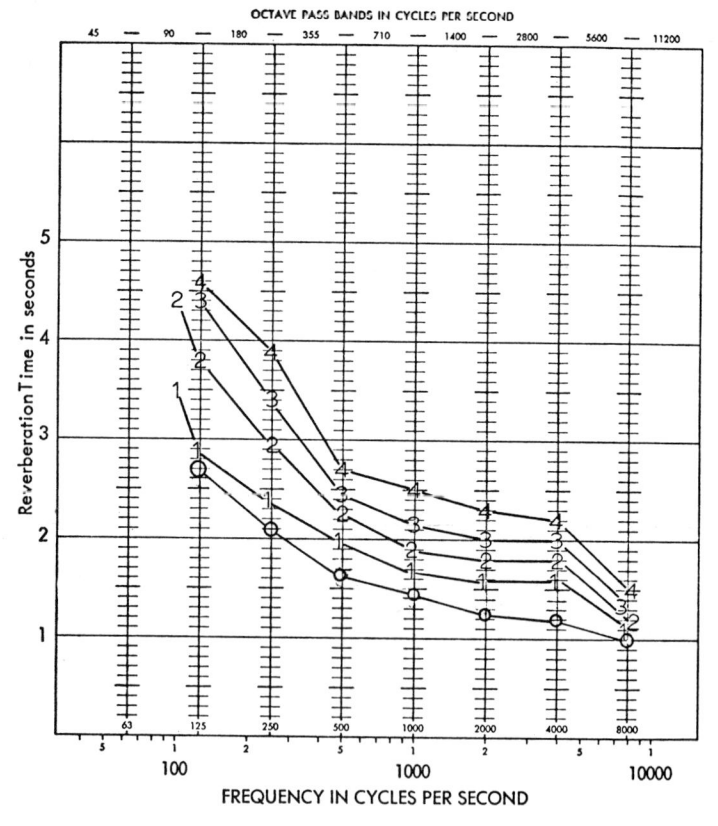

FIG. 17. Reverberation time as a function of stage-to-hall coupling.

Design Considerations of the Professional Consultant 37

The bigger question is—where are we going? For guidance in formulation of criteria we have a different research tool called Auditorium Synthesis. The electro-acoustic system is shown in Fig. 18. The progression of direct, envelopmental and reverberant sound is caused to occur in the small

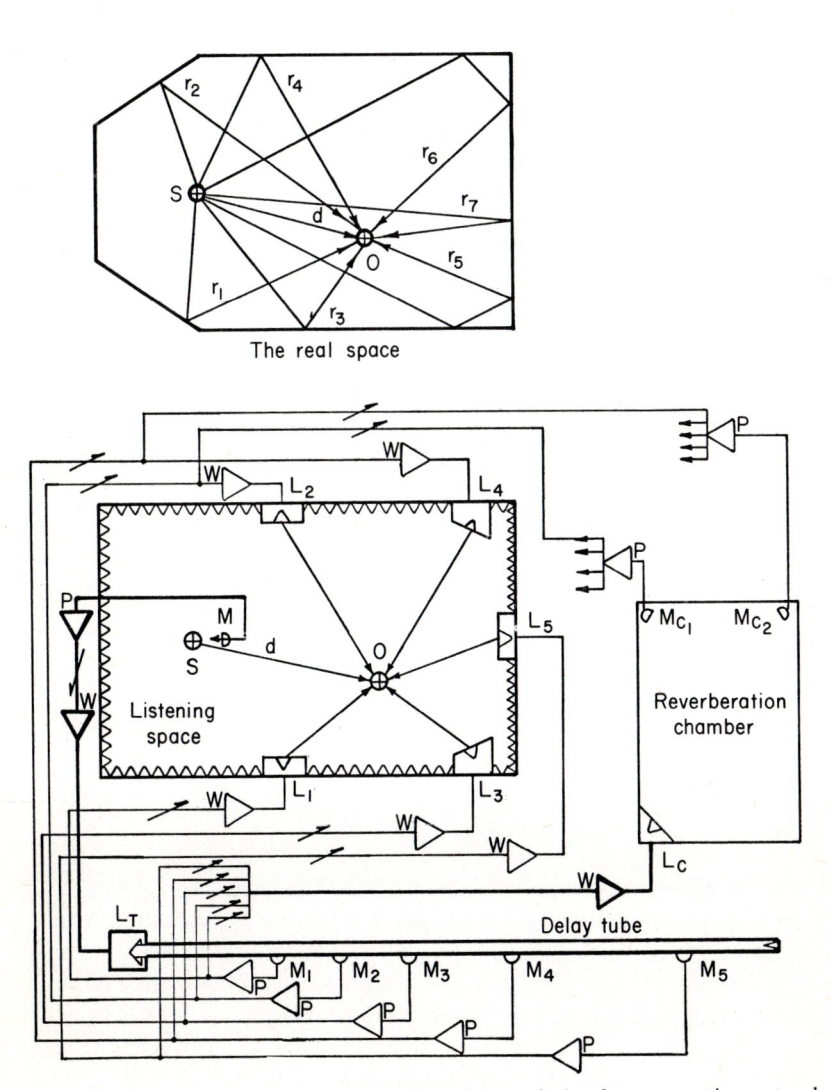

FIG. 18. Auditorium Synthesis—a research tool or a device for augmenting natural reverberation.

studio quite accurately as it does in the full scale hall, as demonstrated by echograms, and all relations can be controlled quite precisely. One of the most important facts first displayed as shown in Fig. 19 was that, as important as are time delays and directions, the *relative levels* of the three key components of auditorium sound are most important. Moreover, the preferred relative levels depend greatly on taste and programme material over a wide range. In particular, the relative level of reverberant to direct sound is far more important than the reverberation time and controls the clarity. Much is to be learned from this research tool. But even more challenging are the problems of translating optimised ratios of the significant variables into real acoustical spaces.

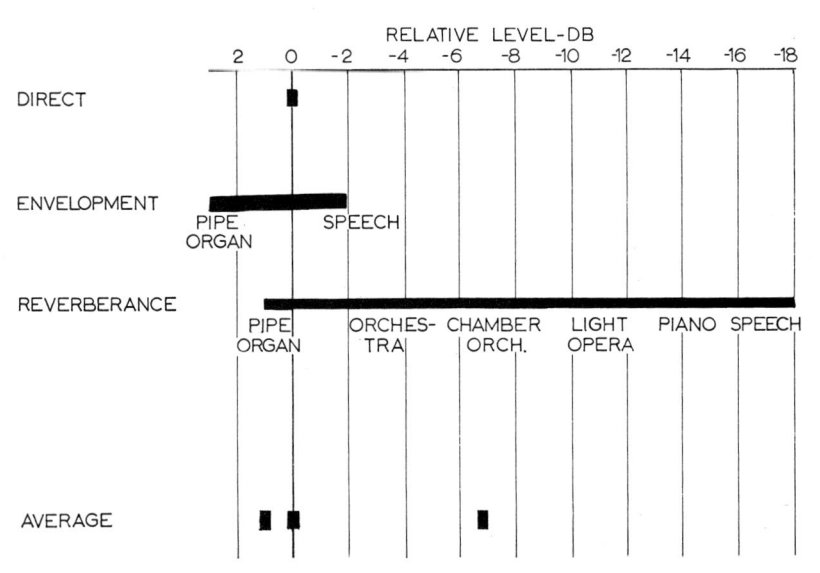

FIG. 19. Criteria for relative levels of Direct/Envelopment/Reverberance sound as derived from Auditorium Synthesis research.

A few more tasks remain for the acoustical designer.

The chairs in an auditorium should stabilise the reverberation time as audience varies. Ideally, the unoccupied seat should equal the absorption of the occupied seat. This has not quite been accomplished. Figure 20 shows our specification. A persistent fault of chair acoustics has been excessive absorption in the range of 250 cps, as shown for a typical sample. This failure can be corrected by proper design.

Although classical music needs no assistance in a good modern hall,

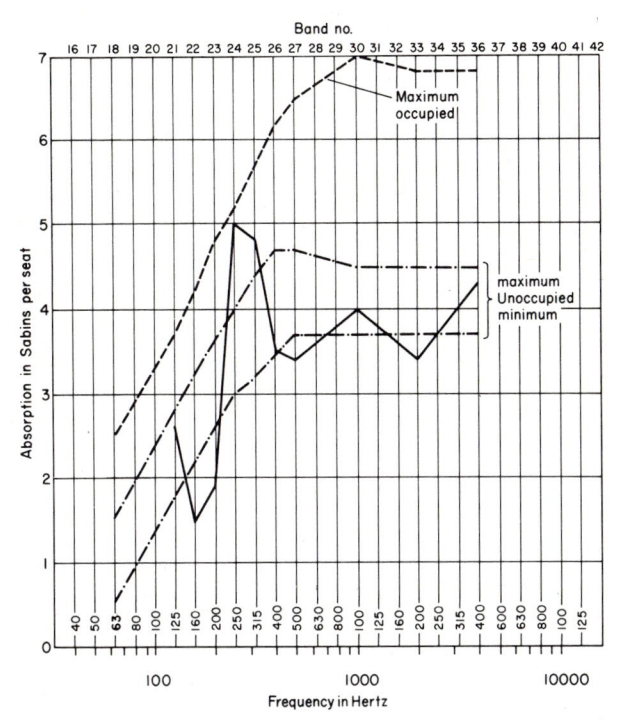

Fig. 20. Criteria for acoustical properties of the theatre chairs and typical faulty performance.

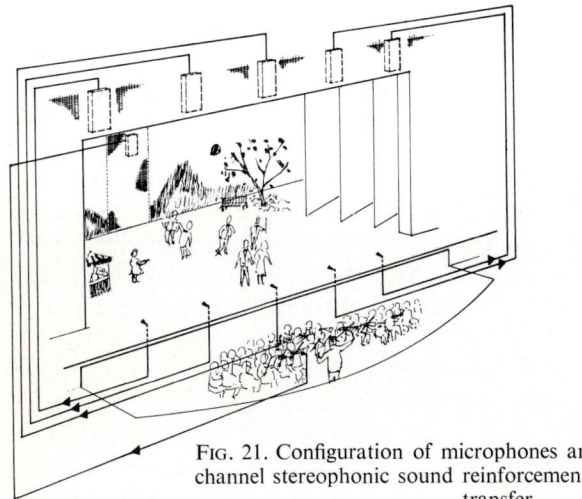

Fig. 21. Configuration of microphones and loudspeakers for five channel stereophonic sound reinforcement plus pit-to-stage sound transfer.

even for capacity as large as 5000, people expect to hear plays and music drama as clearly as they do in motion pictures and television. Discrete sound reinforcement is essential in the larger halls. Figure 21 shows diagrammatically the modern approach. Five pure stereophonic channels give a gentle boost to the direct sound proceeding through the proscenium plane, and preserve precise spacial localisation of sources on the stage. A sixth channel may be used to transfer sound from pit to stage, to help stage players with pitch and rhythm, or it may be used for reinforcement of a soloist. Two more channels may be used for surround sound effects. The loudspeakers are hidden in the architecture of the canopy just above the proscenium. The microphones are inconspicuous on the forestage in special plaques which preserve their free field response and directivity. Thus all appurtenances can be inconspicuous. The result should also be inconspicuous. Ideally, the gain is set and the performance left to the performers. One of the artistic advantages is that the performers can use more artistic dynamic changes.

Pipe organ is a much to be desired feature in the multi-purpose auditorium, but its problems make it a rare feature. In the stagehouse configuration there are many bad ways to install an organ. The best way is the large mobile instrument that can be placed on stage as at the Philadelphia Academy of Music. But this has proved rather impractical. For one thing, stage storage space is precious. We have high hopes for a technique called Organ Transfer. The organ console is mobile on stage. The pipe works are elsewhere in a separate chamber which can be optimised in many ways for the organ. The sound is transferred by three electro-acoustic channels using special mobile loudspeakers. The technique is not new. Modern demonstration has proved it can be done superbly, but we have not been successful in promoting a permanent installation.

The acoustician's work is not finished with the building. He must be prepared to supervise and train local personnel during early performances if the features of a new hall are to be properly used thereafter. For example, a great deal can be accomplished to improve orchestral performance by effective arrangement on stage and, especially in the stagehouse configuration, by selective projection and discard of sound from the stage. Precise model experiments as shown in Fig. 22, with results as shown in Fig. 23, give guidance for the optimal use of riser profiling and ceiling configuration over the orchestra. Another fascinating subject is the internal, and the pit versus stage sound balance for opera which involves details of pit design and acoustical adjustment, as well as the arrangement of the orchestra in the pit. Training in the use of acoustical features and sound reinforcement,

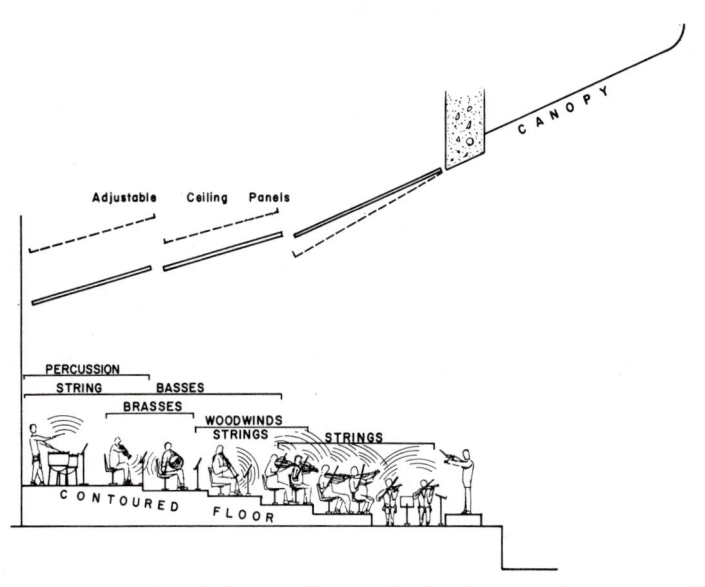

FIG. 22. Illustrating adjustability of ceiling panels of orchestra enclosure over orchestra for partial control of intersection balance.

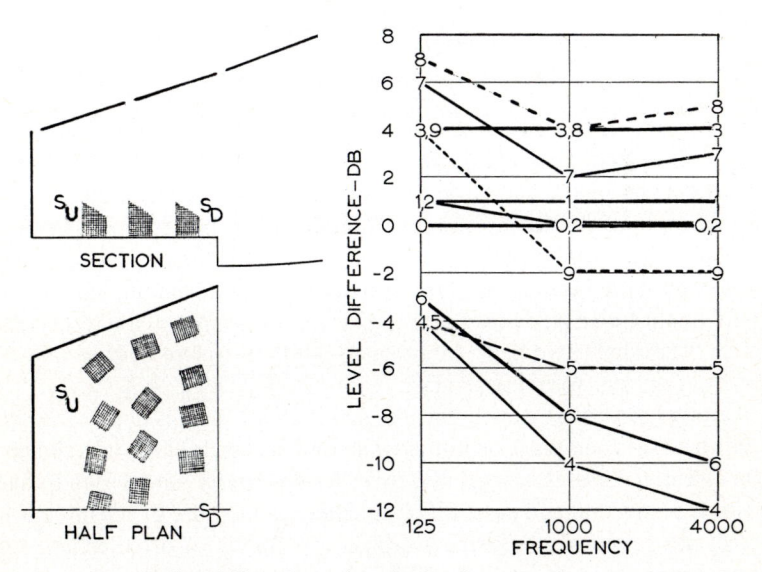

FIG. 23. Range of adjustability of sound projection from sound sources on a stage as a function of location, height, absorption, and ceiling configuration.

and achievement of sympathetic and responsible operation are very difficult and often discouraging. The consultant's task does not end with design if success, *i.e.* the achievement of the potential results of design are to be accomplished.

Finally, there is the task of evaluation of results, which is so vital if we are to build progress upon experience. We now have available a powerful four link chain, illustrated in Fig. 24.

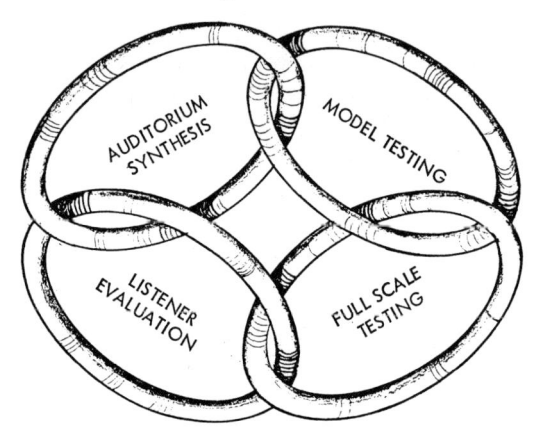

SCIENCE IN THE SERVICE OF
THE PERFORMING ARTS

FIG. 24. The design-evaluation chain.

During the design process we have optical and acoustical modelling for reassurance. Echograms from the models can be correlated with similar measurements in the full scale finished auditorium. Of course, the final evaluation must be the judgments of qualified and perceptive listeners. It is very difficult to translate this information into quantitative guidance, but we must keep trying. And finally to assist this translation, and develop useful quantitative guidance for progress with prescriptions and criteria for better future halls, we have Auditorium Synthesis. Here Science and Art can meet on real speaking terms.

And so I trust that we see that the acoustician's task is detailed design. Cautious evolution must be our slogan: to discover, isolate, test and confirm essential features, to optimise their design criteria; and design to meet them. It seems that acoustics may yet achieve a measure of scientific principles and practical engineering design in support of architecture. Our commitment, mission, and rewarding goal may be Science in the service of the Performing Arts.

5

Multi-purpose Auditoria: an American Phenomenon

Theodore J. Schultz

*Bolt, Beranek and Newman Inc.,
Waltham, Massachusetts, U.S.A.*

R. Lawrence Kirkegaard

*Bolt, Beranek and Newman Inc.,
Downers Grove, Illinois, U.S.A.*

Rarely can an American city, town or university community afford the luxury of building separate facilities for opera, concerts, and dramatic productions; it is simply too difficult to support separate halls. Both the birth and continuing life of a public performing space in America depend upon its ability to accommodate an extremely broad variety of performances, from symphony orchestra to rock concerts, opera, musical comedy, film series and travelogues, lectures and dramatic productions. The acoustical design of such multi-purpose spaces must account for this variety and, to as great an extent as possible, must provide each type of performance with the acoustical environment most suitable to both performer and audience. Even within a single use, such as for symphonic music, it is desirable to provide some adjustability of the acoustics, in order to accommodate both baroque and romantic repertoire satisfactorily.

Understanding what elements and design approaches determine the appropriate acoustical environment for each performance type is critical. Of all the uses mentioned above, opera and concert hall uses are probably the most demanding determinants of gross volume and shape; they are also the most critically judged in terms of final suitability. For the other functions the space is largely a medium for communication, whereas for opera and concert use the space becomes an extension of each voice or instrument.

In an attempt to determine what physical and acoustical characteristics the best halls have in common, Beranek dedicated himself, in the late fifties, to a study of more than 50 concert halls and opera houses. The results of this study were published in 1962 in his book *Music Acoustics and Architecture*.

This book presented rating scales for concert halls and opera houses. In practice, a number of merit points are given for each of a series of objectively determined characteristics of the hall, along with penalties for such defects as echoes or tonal distortion.

The rating scale was so designed that the total score for the hall correlates very well with the consensus of subjective judgments about the acoustical quality of the hall, as expressed by musicians, conductors and critics.

The trouble was that these rating scales are of little help in guiding the design of a new hall, except that they do establish certain priorities that indicate where to concentrate the design effort.

For example, Beranek's notion of the 'initial time delay gap' was new, and the results of his studies led him to assign great importance to achieving the proper value for this quantity: it influences the rating much more strongly than any other factor. So far, so good! It is easily possible to provide for early reflections at the right delay following the direct sound (which is what the initial time delay gap is all about); but nothing is said in the rating scheme about the strength of these early reflections, nor about where they should come from. Other matters, such as choosing and achieving a low background noise level, a suitable reverberation time, a suitable balance between low-frequency and high-frequency reverberation, for example, are amenable to quantitative prediction; but there is no way to quantify diffusion, balance of instruments, blend of sound, orchestral ensemble, echoes or tonal distortion. Dealing with these questions in designing a hall was still very much a matter of experience, and not much help was offered by the rating scales.

Looking for more concrete guidance, one can conclude from Beranek's study that the acoustical designer of a concert hall starts with many points in his favour if he simply gives the hall the gross shape of a rectangular 'shoebox'. Of the six halls that were given the highest rating $(A+)$ by conductors, critics and musicians, four were 'rectangular shoeboxes':

	Score (out of a possible 100 points)
Boston, Symphony Hall	96
Vienna, Musikvereinssaal	94
Basel, Stadt-Casino	93
Amsterdam, Concertgebouw	91

The fact that the other two '$A+$' halls were far from rectangular shoeboxes indicates that, though this shape may be sufficient, it is not absolutely necessary for the achievement of fine acoustics:

	Score (out of a possible 100 points)
Buenos Aires, Teatro Colon (Horseshoe opera-type)	95
Lenox, Mass., Tanglewood Music Shed (Low, flat fanshape)	92

Of the halls rated 'A' by the musical experts, five were rectangular shoeboxes:

	Score (out of a possible 100 points)
Zurich, Grosser Tonhallesaal	89
Glasgow, St. Andrews Hall	88
Bristol, Colston Hall	86
Leipzig, Neues Gewandhaus	84
La Chaux-de-Fonds, Salle Musica	92

However, 14 other A-rated halls were *not* rectangular shoeboxes. Only one hall with rectangular shoebox shape was not rated $A+$ or A!

What are the desirable features of the rectangular shoebox hall? Let us

FIG. 1. Boston Symphony Hall—view from stage.

look at the four halls rated '$A+$' by the professionals. Boston Symphony Hall, which scored the highest in Beranek's interviews, is shown in a view toward the rear in Fig. 1. Figure 2 gives the plan drawing and longitudinal section.

Note the absence of a proscenium, and thus no 'doghouse' for the orchestra, such as frequently appears in auditoria that must also have a stagehouse for productions requiring scenery. The performers and the audience in a shoebox hall are essentially occupying the same large room.

Note also the narrowness of the hall: it is only about 75 ft wide; moreover, the balconies typically extend along the entire sidewalls up to the stage, and the distance between balcony faces is of the order of only 60 ft. There are many diffusing surfaces that scatter the reflected sound in all directions. And there is minimal balcony overhang, at least at the sides of the hall. The balcony soffits 'catch' the early sound reflections from the sidewalls and turn it back down toward the middle of the main floor seating area. It is this area, incidentally, where the most serious acoustical prob-

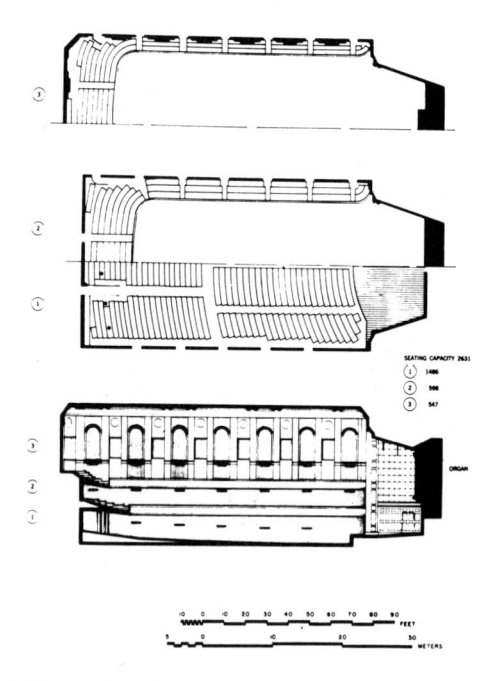

FIG. 2. Boston Symphony Hall—plan and sectional elevation.

lems tend to show up, particularly in the large halls needed to house the large audiences needed to support the large orchestras of today. This area is often starved for reflections arriving early enough to clarify the rather muddy sound that is typical there.

With certain variations these features can also be seen in Vienna's Grossemusikvereinssaal (Figs. 3 and 4), Basel's Stadt Casino (Figs. 5, 6 and 7), and Amsterdam's Concertgebouw (Figs. 8 and 9).

These simple architectural ground rules, apparently, were sufficient to assure almost automatically the achievement of those acoustical characteristics that the research of the sixties has shown to be desirable in concert halls: 1) the proper temporal pattern of arrival for early sound; 2) the proper space and time distribution of sound arriving from different directions; 3) the sense of proper 'surround', and other acoustical virtues—provided that the hall dimensions were not too great.

These ground rules were, of course, supplemented with purely acoustical criteria, mostly concerned with reverberation time and the avoidance of foci and echoes, to form the basis for the 'pure' concert hall design philosophy.

Although, as noted above, other shapes than that of the rectangular shoebox can be made to work excellently, the acoustical benefits of the rectangular hall seemed to be a firm basis for discussing preliminary design with concert hall architects; we particularly urged the avoidance of novel

FIG. 3. Vienna Grossemusikvereinssaal—view towards stage.

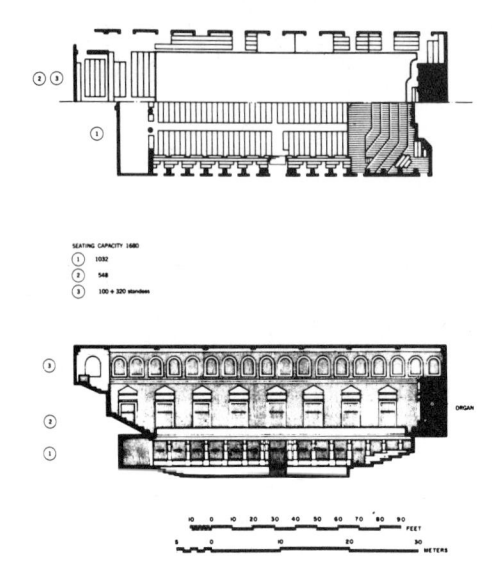

FIG. 4. Vienna Grossemusikvereinssaal—plan and sectional elevation.

FIG. 5. Stadt Casino Basel—view towards stage.

FIG. 6. Stadt Casino Basel—view towards rear.

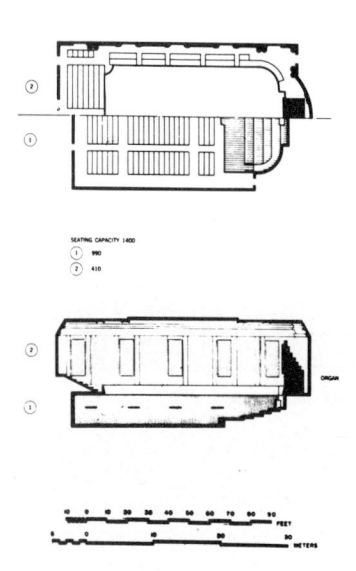

FIG. 7. Stadt Casino Basel—plan and sectional elevation.

FIG. 8. Amsterdam Concertgebouw—view towards the rear.

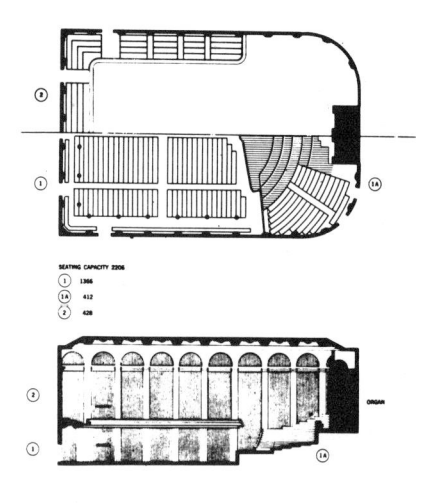

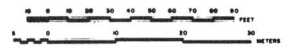

FIG. 9. Amsterdam Concertgebouw—plan and sectional elevation.

shapes for new halls. This concept formed an important part of our auditorium work in the early sixties.

In Bolt Beranek and Newman's current work, that design philosophy has been adapted and refined in order to provide excellence in concert hall acoustics while meeting all the physical, acoustical and theatrical provisions required by the other performance types. The necessary elements in this design approach include:

(1) Sufficient volume for maximum occupied reverberation times of 2·0 to 2·4 seconds at mid-frequencies. The modern practice of recording symphonic music in a reverberant hall with close microphones has developed a preference among classical music audiences for sound that is both quite live and quite clear, a combination that is difficult to achieve in a concert hall!

(2) Critically placed, broad-band adjustable absorption to reduce mid-frequency reverberation times to 1·4 to 1·6 seconds, suitable for musical comedy, opera, and even drama. This great a change in reverberation time requires that a very large amount of absorptive material be added to or removed from the room: an area comparable to that occupied by the audience. It is usually a challenge to find space for this material and a way to hide it away when it is needed.

(3) Seating capacity limited to 2400 seats. Although it is possible to design larger halls with fine acoustics, the problems mount rapidly with size increase above this seating, and more of the architect's necessary compromises must be forced in favour of the acoustics. He loses flexibility in his visual design choices.

(4) Full theatrical rigging capability. Facilities for staged productions must not be skimped; but the lighting requirements and the need for high quality sound reinforcement and sound effects systems introduce great conflicts with acoustical needs that are not confined to the neighbourhood of the stage.

(5) Excellent sight lines and close proximity of audience to stage. Modern audiences will not be satisfied without roomy seats, and when a large audience must be accommodated, the greatest hazard of the narrow rectangular shape quickly becomes apparent: too many of the audience are simply too far from the stage to feel involved in any way with the performance.

(6) Concert enclosure and front-of-audience-chamber shaping to approximate 'pure' concert hall configurations; audience chamber portions able to be reduced in width, depth and height to optimise space/time distribution for speech-related performances, and to provide an

audience chamber of appropriate shape and size for the type of performance, in each case.
(7) 'Silent' air distribution systems. Background noise levels near the threshold of audibility are necessary in order that the audience may fully enjoy the occasional breathless moment that makes a great musical performance into a living memory.
(8) Electronic sound reinforcement and flexible theatrical sound effects systems.
(9) Mechanised orchestra pit lift with three operating positions:
 (a) orchestra pit level height (adjustable for conductor)
 (b) extended stage area
 (c) additional audience seating when (a) or (b) are not required
 (d) if the programme for the hall features frequent use for opera, the pit elevator should be divided into two sections, each running about the full width of the stage. For greater flexibility, each elevator section should have two floors, separated about 8 ft in elevation; with the upstage elevator in its upmost position, the upper floor forms an extension of the stage and the lower floor is the upstage half of the pit floor; the downstage elevator section might then be in its lowest position so that its upper floor forms the front part of the pit floor. Alternatively, with the front section in its upmost position, its upper floor forms a cross walk parallel to but separated from the stage, with a portion of the pit orchestra seated on its lower floor. Such installations have been found very flexible and have inspired ingenious innovations in production.

What are the conflicts in relating these design approaches to those that would apply to the 'pure' concert hall, and how can they be resolved? What are the subjective/objective results that have been achieved? And finally what directions do we see future designs taking? These questions were addressed, in the paper as presented at the Edinburgh symposium, and were illustrated through numerous photographic colour slides of completed projects and of work-in-progress. The illustrative material in the lecture was presented rapidly; it covered some of the most interesting multi-purpose halls with which we have been involved in the last fifteen years. The audience was not expected to take in all the details of the many halls that were presented; but we hoped that, by a sort of 'persistence of vision' they would come to recognise a common thread of design philosophy behind the various halls, stemming from the natural characteristics of the rectangular shoebox shape. A secondary purpose behind the multitude of illustrations was to present a variety of concrete suggestions for how to realise in three

Multi-purpose Auditoria: an American Phenomenon 53

dimensions the concepts that, as acousticians, we often can only talk about.

It will be interesting, in conclusion, to illustrate our current design approach by discussing a new concert hall, now in the planning stages, for Melbourne, Australia, as shown in Figs. 10 and 11; it will seat about 2700 people.

Because of the new directions being taken by composers of music, the concept of the concert hall must be stretched nowadays to the point where it becomes, itself, practically a multi-purpose facility. It requires, of course, a full (and large!) stage for orchestra. But also many conductors want much

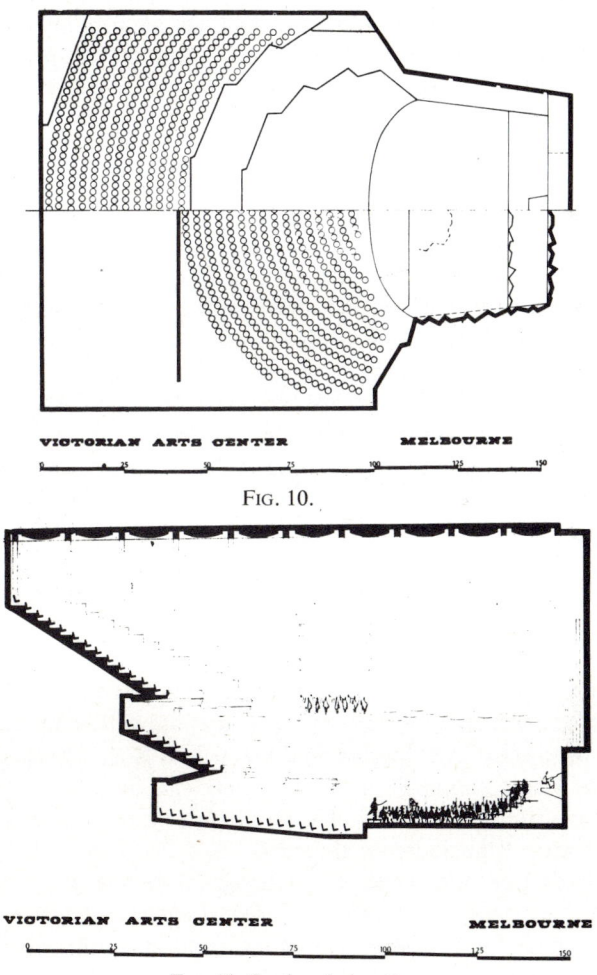

Fig. 10.

Fig. 11. Sectional elevation.

greater flexibility in disposing their musical resources than in the past. Accordingly, even in this 'pure' concert hall, there will be an orchestra pit, with divided elevators, to accommodate about 60 musicians.

There is an organ, permanently located on the rear wall, and elaborate provisions for chorus, including an upstage elevator that can carry either of two sets of seating: one with close spacing for the chorus, the other with more spacious allowance for audience. When used for chorus, the seats may rake up from the stage floor behind the orchestra, in the usual manner; or they may be elevated to match the balcony seating that extends alongside the stage. For that case an alternate rear stage wall is provided on the front of the elevator below the chorus seats. When no chorus is required, the alternate seating wagon is moved onto the chorus elevator, and it is raised to continue the balcony seating behind the orchestra. The unused seating is stored under the stage.

There are performer locations at a number of different places about the hall; there are projection facilities and provisions for 'surround sound', as well as the more usual sound reinforcement and reproduction systems.

But the greatest departure is the provision of extreme visual intimacy, with as much as possible of the audience quite close to the performers. This results in a hall that, in order also to achieve suitably high reverberation time, is almost cubical. We have struggled for years to fit large audiences into the narrow shoebox shape, and have invariably discovered that the hall must end by being considerably wider than we want, anyway; early sound reflections had to be introduced from an overhead position. If we admit this from the start and introduce the element of visual intimacy, allowing the hall to become as broad as it needs to, we find that we enter an unexpected domain of acoustical advantage.

Note that the first row of the first balcony is only 42 ft from the stage and that the most distant seat is only 100 ft away; 75% of the seats are within 75 ft of the stage. As a result, many of the balcony seats are in the direct sound field of the orchestral instruments that need most to be aided by early reflections. Accordingly, only minimal use will be made of overhead sound reflectors serving the mid-main floor area; the lacework of reflectors will be combined with a chandelier spanning the entire stage. There will be extensive banners of variable absorption on the side walls, toward the rear of the audience chamber.

It is expected that the extremely good sight lines and great visual and acoustical intimacy achieved through closeness of the audience to the performers will combine with excellent acoustics to make this a worthy extension of the shoebox concept.

6

Model Studies with Particular Reference to the Sydney Opera House—the Evaluation of Objective Tests of 'Acoustics' of Models and Halls

Vilhelm Lassen Jordan

Gevninge, Dk 4000 Roskilde, Denmark.

In the first part of this paper I would like to give you an indication of the design considerations of the Sydney Opera House with special reference to the model studies and following that I am grateful to have the opportunity of discussing one of my favourite ideas and concepts of acoustic criteria.

The story of the Sydney Opera House goes back to the competition in 1956 won by the Danish architect Jorn Utzon and to the actual construction which started in 1959 giving a 14 year period between the start and the completion in 1973. During this delay, unheard of in the history of most auditoria, there was a change of architect which is also rather unusual, so making this quite a unique consideration in the whole development of the acoustical ideas. I would like to illustrate this with some illustrations of the very first concepts, followed by some from the intermediate stages and finally some of the actual designs as executed, after which I will give you an idea of the model studies performed.

Figure 1 illustrates the first concept of the so called major hall. At the time it was thought of as a concert hall which could be converted into an opera so that the concert hall had 2800 seats and the opera only 1800 and this was to be made feasible by putting the audience right on the stage, closing the whole stage loft over with solid panels and thus forming a concert hall. Now this is to me today absolutely outdated and traditional in the way it thinks of sound distribution from the ceiling for instance. A cross section at the rear of this concert hall is shown in Fig. 2 and there is a funny thing about this: at the time nobody really realised that these vertical walls could not be housed within the shells and that the design was completely unrealistic. Figure 3 shows the plan of the concert hall and you

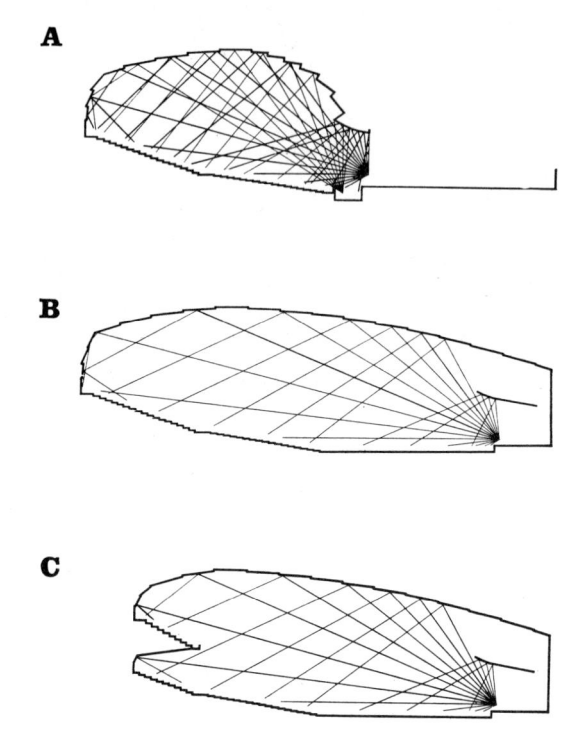

FIG. 1. First concepts of the Major Hall—longitudinal sections (A) Grand Opera (B) Concert Hall without balcony (C) Concert Hall alternative with balcony.

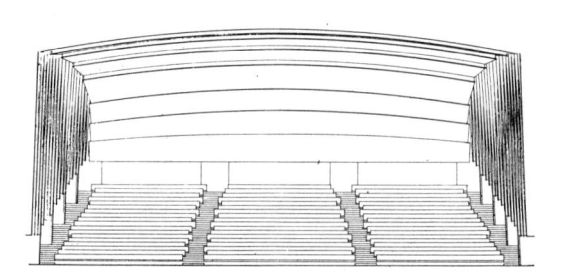

FIG. 2. Cross section through the Concert Hall (first concept).

can see the oblong shape, the immense area to be covered and the stage in the far end.

Then we came to the next concept and it is still the Utzon idea but now they have all realised that to house this inside the shells, you cannot have the vertical side-walls as they will have to bend inwards. Utzon invented a beautiful diamond ceiling (shown in Fig. 4) which would give a lot of

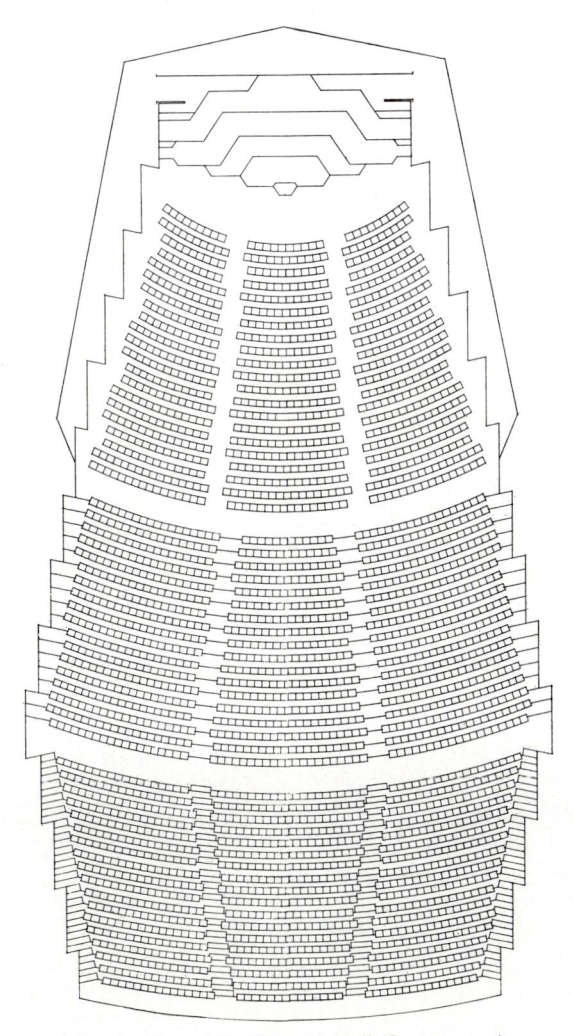

FIG. 3. Plan of the Concert Hall (first concept).

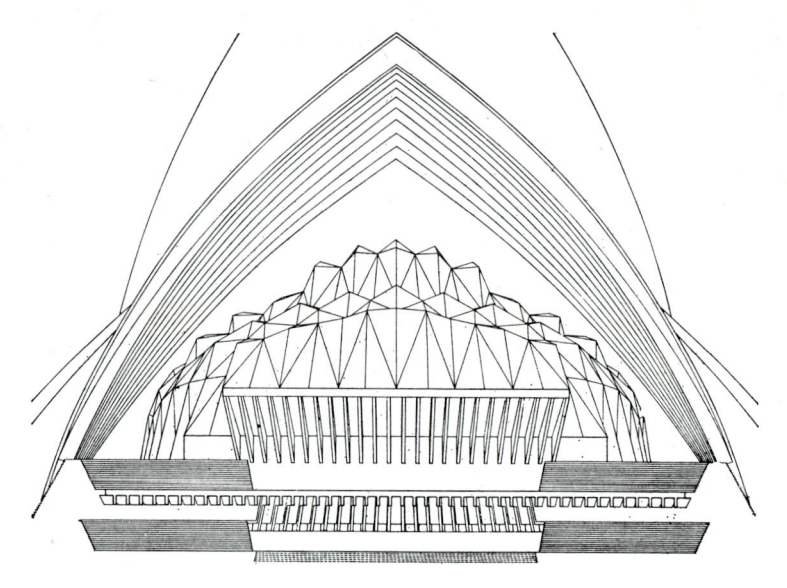

Fig. 4. Diamond shaped ceiling as proposed in second design of Major Hall.

Fig. 5. Second concept of the Minor Hall—longitudinal section.

diffusion but unfortunately, or maybe fortunately, this concept also was changed and we are now approaching the final stage which shows a second concept of the minor hall (Fig. 5). The minor hall was to be the drama theatre for 1200 people housed under the second sized shell and you see here, this is a very, very funny shape because it has some sort of vault-like structure in the middle of the auditorium and I must say that the architect and myself, we were not exactly friends when he had this idea since I realised of course the acoustical problems we had to face. Well, luckily he went away from that one. Figure 6 once again shows the major hall and this is the third concept of Utzon. By this time he had given up the idea of using the stage proper for seating the audience and had drawn out the audience in front of the proscenium frame losing lots of seats, although

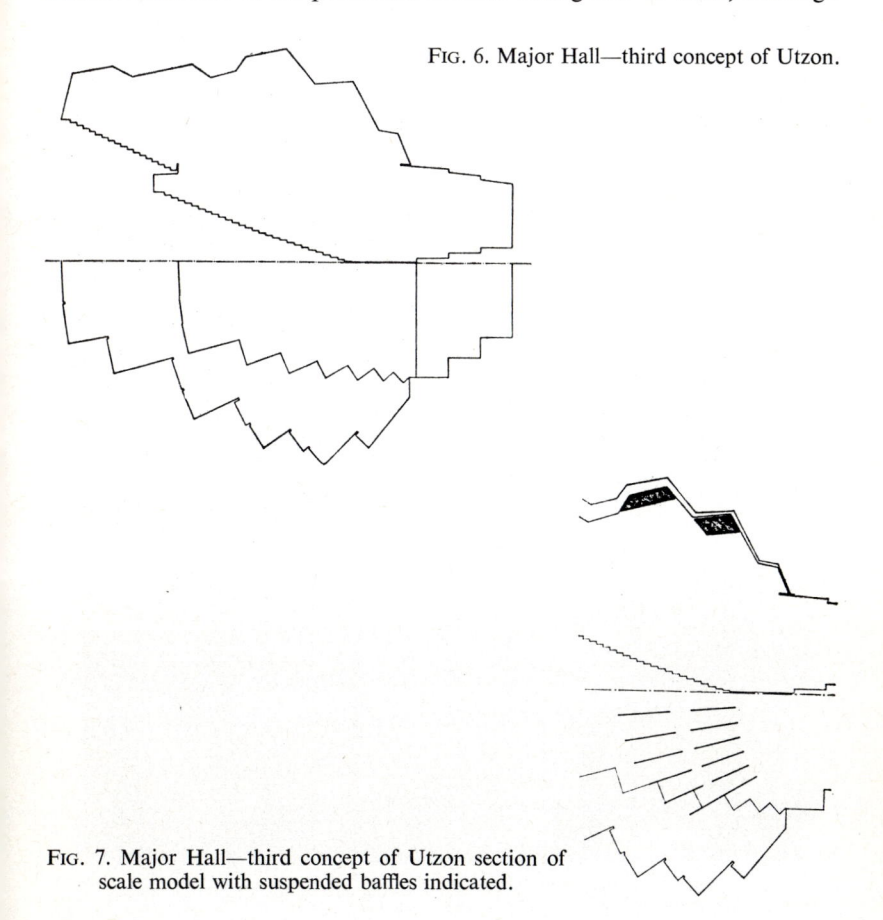

Fig. 6. Major Hall—third concept of Utzon.

Fig. 7. Major Hall—third concept of Utzon section of scale model with suspended baffles indicated.

he never really disclosed how many. He also used a smaller reflector for the orchestra. Figure 7 shows a section through the model which we made at that time for the major hall. The ceiling had a very polygonal shape and one can see that it was a very wide hall, so we gave some thought at the time to ways of getting some reflections to-and-fro sideways. The only place we could put reflectors to give sideways reflections was high up where we put some baffles, which can also be seen in the section. This was tried out on the model and it showed a definite improvement in the criterion we used at the time.

Next we come to the new phase, the phase which was realised. The change of architects led to quite another concept for the building. The idea of the combination of a concert hall and an opera was abolished and it was decided that the major hall should be solely a concert hall. Now the consequence of this was a complete redesign of the interior in which the stage was moved forward a little so that the audience is sitting around. In Figure 8 we are looking across the orchestra stage and one can see the baffles hanging from the ceiling. This is the model in $\frac{1}{10}$ scale and you see the choir seating and also audience seating behind the orchestra stage which is at the lower level of the picture. Figure 9 shows a view towards the rear, where most of the people are sitting in the stalls and in the 1st and 2nd terrace with some at the side boxes. I mention the side walls because this problem occurred since the shells slope in—you cannot do anything about it. If you

FIG. 8. Model of Major Hall—shown looking across the orchestra stage.

want some vertical sides to give you some side reflections, what do you do? The first idea was to step them, as you see they are stepped from the side inwards. Now in the ceiling, the idea was to have some sort of catenary shape, which is very good to distribute the sound but on the other hand we discovered by the model research that this particular ceiling with these very big catenaries meant a so-diffuse and so-dominating sound from the ceiling, that the side reflections were more or less lost. Figure 10 shows a plan of

FIG. 9. Model of Major Hall—shown looking towards the rear.

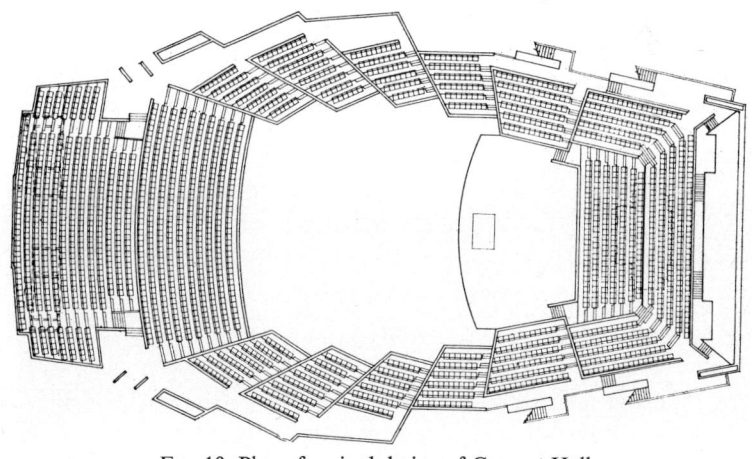

FIG. 10. Plan of revised design of Concert Hall.

the new design and you see the stage, the choir seating, the seating around the stage, the side boxes and the 1st and 2nd terrace. Figure 11 shows the model of the second design. The idea of this change from the first design was first of all to develop vertical sidewalls. This was made possible by drawing in the side walls so that the side boxes were brought a little under the soffits and you can see that quite a considerable height of side-wall is possible with this shape. Figure 12 is looking the other way in the model.

FIG. 11. Model of second design of Concert Hall—looking across orchestra stage.

FIG. 12. Model of second design of Concert Hall—looking towards rear.

This model had a scale of 1 to 10 and was tested with spark gap and a ¼ in B & K microphone and you can see the side-walls which are diminishing at the rear. Figure 13 gives a good idea of how this was built into the shape under the shells. I am sorry to say, because architects always ask this question, 'Has the interior something to do with the exterior?' and I must say quite frankly, nothing at all. On the other hand I must admit, judging the whole story of this change of architect I believe that this combination has a very fine artistic exterior with a more rational and planned interior, maybe just the optimal solution. The former minor hall shown in Figs. 14(a) and 14(b), now called the opera theatre, was increased to 1500 seats with balconies and side balconies. There is one set of features in the plan which I would like to point out to you because I believe they are important for the sound in the theatre. These are the side splays which are actually protruding a little into the stage itself causing the voices of the singers not only to be reflected and carried out into the auditorium but also to be sent back to the singers themselves. Figures 14(a) and 14(b) show the first design of this opera theatre and this had a rather flat sloping ceiling. The design was found by model research to be not so good and in addition the model test showed that the ceiling reflection was too dominant in comparison with the side reflection, so that in the final shape the ceiling weight was increased as much as possible and the ceiling pushed, so to speak, up against the underside of the shells. Figure 15 shows the view from the stage of the

FIG. 13. Concert Hall as part of total design: (1) Concert Hall; (2) Rehearsal/Recording Hall; (3) Drama Theatre; (4) Drama Theatre Stage; (5) Production Rehearsal Room; (6) Public Lounge; (7) Admission Offices; (8) Admission Offices; (9) Control Room for Rehearsal/Recording Hall; (10) Foyer; (11) Exhibition Hall and Chamber Music Hall/Cinema; (12) Car Concourse; (13) Restaurant.

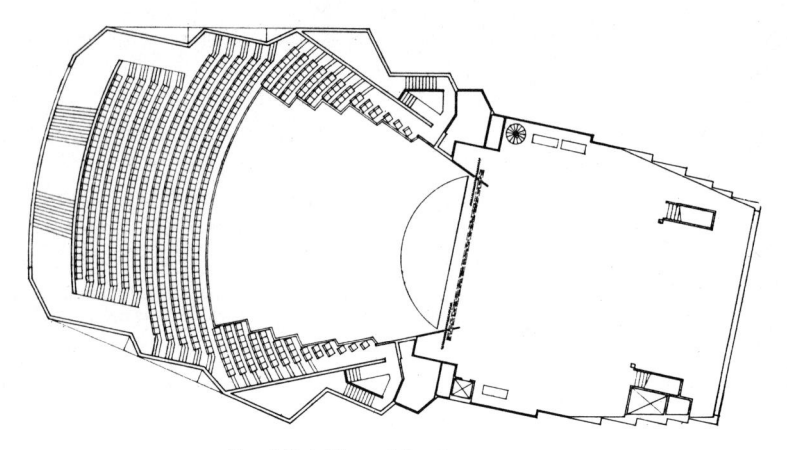

Fig. 14(a) Plan of the Opera Theatre.

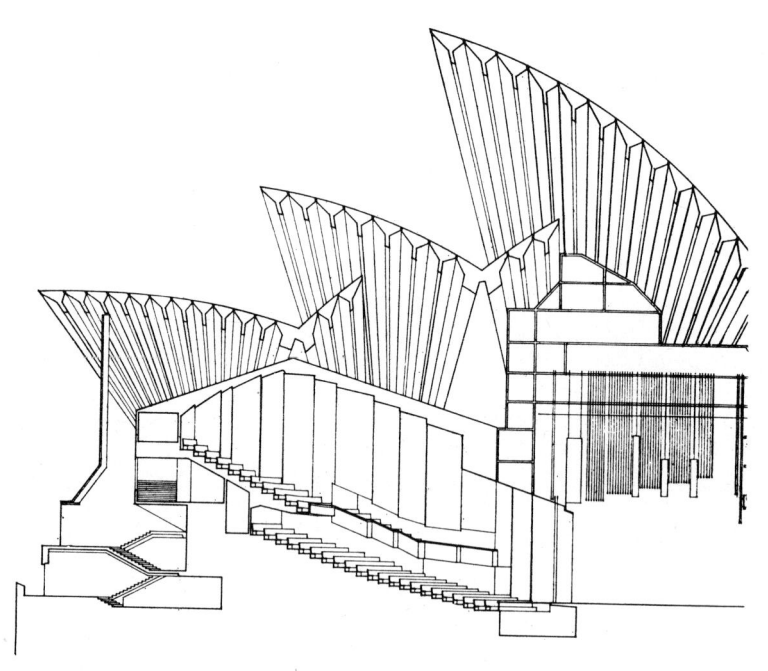

Fig. 14(b) Sectional elevation through Opera Theatre.

model towards the audience and you can see the side balconies and also the spark gap which was used for these tests.

In the original concept there was an immense understage for the grand opera in the major hall, which meant that there was a space left when the decision to make the major hall into a 100% concert hall was taken. This understage which was left over was developed into an orchestra studio of quite significant volume—between five and six thousand cubic metres and the reverberation time can be seen from Fig. 16. I adhere to the philosophy of not putting too much attention on reverberation time as a criterion, with

Fig. 15. Model of Opera Theatre.

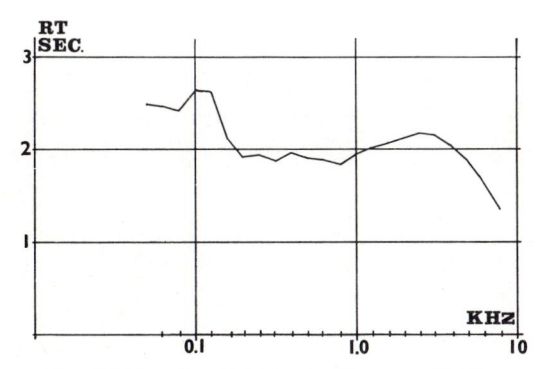

Fig. 16. Reverberation time in Orchestra Studio.

one exception: I still believe that the frequency dependency of reverberation has some bearing upon what you hear and personally I favour a curve of this kind. Not necessarily so much the increase at the low frequencies—I know there are people who favour that very much—but the increase in the higher frequencies as far as you can go and meaning that there should be a sort of saddle in between the lows and the highs.

The noise level which is shown in Fig. 17, was measured inside, from a helicopter at some 200 feet above sea level and was found to be practically or mostly inaudible. Figure 18 gives the sound insulation between the two most important places—the concert hall and the rehearsal or orchestra studio below—and it is quite good as you see. The directivity pattern of the loudspeaker columns which are specially shaped for such an auditorium are given in Fig. 19. They are 4 m columns and they can produce directive sound which is good enough to make the concert hall into a congress hall with a reasonable sound level.

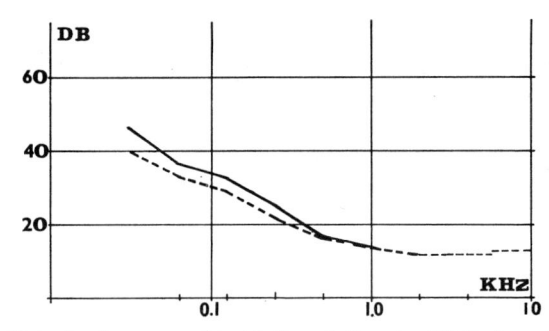

FIG. 17. Noise levels measured inside from helicopter 100 ft above the shells.

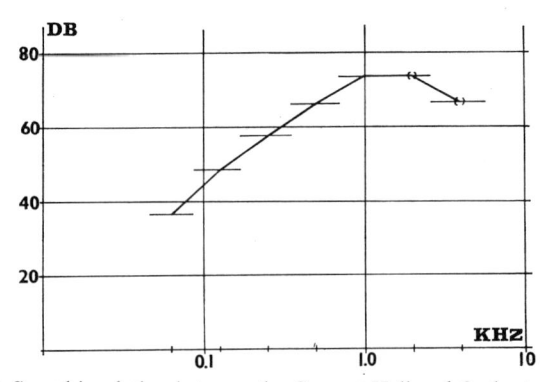

FIG. 18. Sound insulation between the Concert Hall and Orchestra Studio.

The finished concert hall is shown in Fig. 20 whilst Fig. 21 gives the final reverberation time of the hall with and without the audience. In the opera theatre, we deliberately made the seats of leather so that we could

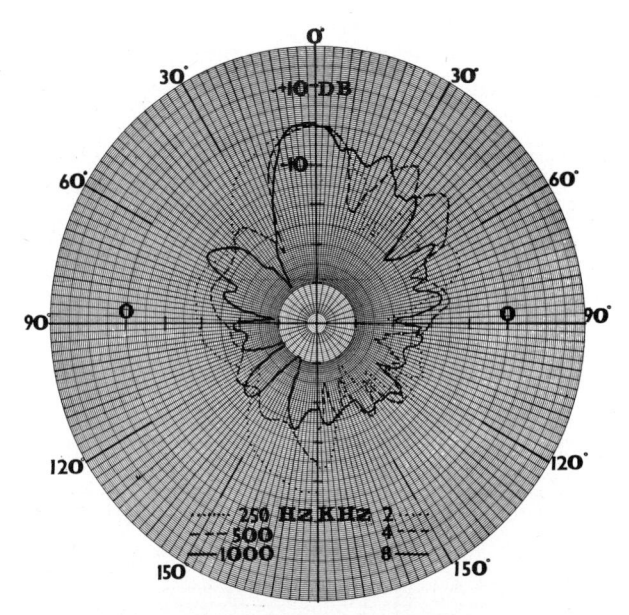

FIG. 19. Directivity pattern of the loudspeaker columns.

FIG. 20. The finished Concert Hall.

have the benefit of adding some reverberation when they were empty but the final result of 1·4 seconds (Fig. 22) was even a little more than we expected.

If I may now turn to the second theme—the evaluation of objective tests of acoustics of models and halls. For some years I have favoured this concept of early decay time which really just means the slope of the first 10 db of the reverberation curve or the decay curve. I have tried to relate this in some way to the shape of a hall by measuring the early decay time in various locations and I have suggested a measure which could give you an idea of what I call the inversion of an auditorium. Now if you measure the

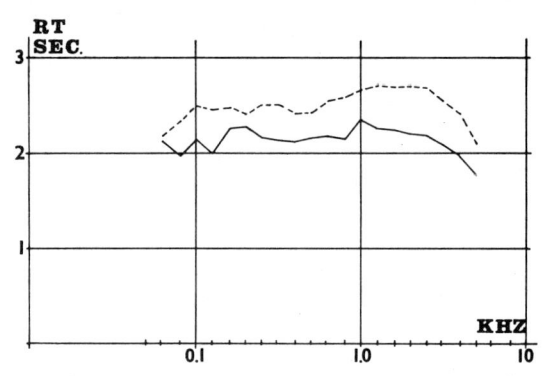

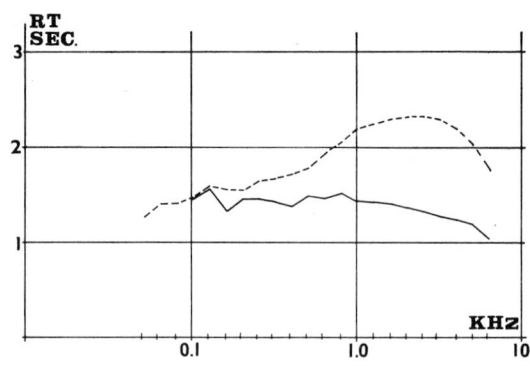

FIG. 21. The reverberation time in the Concert Hall.
────── capacity audience   ----- empty

FIG. 22. The reverberation time in the Opera Theatre.
────── capacity audience   ----- empty

early decay time at a number of places on the stage and at a number of places in the auditorium and you take these two averages and compare, the theory is that early decay time on the stage ought to be shorter than in the auditorium. Why is that? Yes, because you want the sound fields to build up more rapidly on the stage to the benefit of the musicians and the conductor. So if you define the inversion index as the average of the early decay time in the auditorium versus the same figure on the stage, this coefficient should at least be 1·0 or maybe greater. Now this has been tried out in a number of cases recently, first in the models of the Sydney Opera House and then in the full-scale halls measured with gun shots which the audience had to suffer. I want to show the results, but before I do so I would like to point out two other methods of using this the early decay time (E.D.T.). You can compare the E.D.T. directly with reverberation time saying that at least the early decay time should be the same as reverberation time. You could allow maybe 10% less than reverberation time meaning that you would acknowledge all values above 90% as good values, this

TABLE 1. Revised Comparison of Concert Hall and Model Criteria

Criterion & Conditions	Model (2–16kHz)	Hall (250–2000 Hz)	
		empty	cap. audience
Inversion Index (Ideal: ⩾1·0)			
Reflectors at soffit level	1·11	1·14	0·96
Reflectors just below crown	1·05	1·10	1·06
Reflectors 34 ft above stage	—	—	1·05*
Percentage 'good' E.D.T. values (Ideal: = 100%)			
Reflectors at soffit level	94	96	85
Reflectors just below crown	83	100	85
Reflectors 34 ft above stage	—	—	94*
Percentage overall variation of E.D.T. values (Ideal: = 0%)			
Reflectors at soffit level	20	13	17·5
Reflectors just below crown	21	13·5	17·5
Reflectors 34 ft above stage	—	—	10*

* Means frequency range only 500–2000 Hz

is in the audience area. This is the second method of using this E.D.T. concept. The third method is just finding out how much it does vary over all, that is the difference between the highest and the lowest of these figures as a percentage figure throughout a hall and then conclude that the least variation is the best, and gives the most homogeneous hall. We have tried this concept out in all three cases, in the model and in the full scale hall and I want to show you in Table 1 the results from the concert hall at the Sydney Opera House.

The first says inversion index and there are two figures. I did not actually show the reflectors but there are reflectors above the stage in the concert hall and they can be put at several positions and in the model they were mostly tried in a position which corresponded to the ceiling height of the soffits (above the side boxes). If you put the reflectors on the same level you get the result as shown, the reflectors at soffit level 1·11, this is the inversion index. If you pull them all the way up to the ceiling there is only a small variation to 1·05 so the verdict in both cases is valid.

Now repeating that in the full-scale hall you can compare the figures directly and you see a funny thing,—the empty hall compares fairly well with the model but for the hall with capacity audience the inversion index goes below 1—which is bad, or which is considered bad. If you position the reflectors just below the crown, you get about 1·06. This is notable because the new frequency range is only measured from 500–2000 in the third case but it does not really matter because the change with frequency is very little. Now to go to the next criterion good percentage E.D.T. values means values above 90% of the reverberation time and you can see that with reflectors at soffit level this gives 94% whereas just below the crown you have 83%, so there is the benefit of reflectors, but not so in the hall itself where it does not show really. The funny thing is that we tried an intermediate stage at one of the first test concerts with the reflector positioned 34 ft above stage level and it came out better than any of the others so I advised that the reflectors be kept there and they have since been kept at this level. Now to take the third criterion that is the average percentage of all variations of E.D.T. values which more or less indicate how homogeneous your sound field is. Now here you see an interesting point, the model is not diffusive enough. The variation in the model throughout is more pronounced than in the real hall, of course, and the empty hall is less than the full hall, which is natural, but once again it is interesting to note that the level 34 feet comes out best for the reflectors and the variation drops down to 10% so at least in this case the methods seem more or less to coincide.

The values in the opera theatre, that is the smaller hall, are shown in Table 2 and there is of course a big difference when you have the empty hall: the inversion index goes up because of this increase in reverberation time but if you compare the model with the capacity audience hall, you see the figures are very close to each other indeed. Once again the model is much less diffuse than the real hall which can be seen from the figures in the last row. So I would like to query this whole thing because I am the only person who consistently has been using this E.D.T. for a number of years and I am really anxious that somebody should take up this field because before it is confirmed and before it is tried out by other workers I do not consider it absolutely valid. On the other hand I feel that there is something in it which gives not only a criterion, but gives us also some hints of what the shape of an auditorium means. This inversion index is related to the shape and is also related to the amount of side reflections. We have been able in another model to show this by using a directional microphone (using a $\frac{1}{2}$ in B & K put into a parabolic reflector) and we measured the directivity, and at 8 kHz and 16 kHz octaves, which are the usual frequencies for the model the arrangement is very highly directive. So we can sort out the E.D.T. coming from the side versus the E.D.T. coming from above or from the stage and compare. It shows that the side-walls give values of E.D.T. which come close to the statistical value which means that this is

TABLE 2. Comparison of Opera Theatre and Model Criteria

	Model		Hall			
			empty		capacity audience	
Source at	Pit	Stage	Pit	Stage	Pit	Stage
Criterion						
Inversion Index (Ideal: $\geqslant 1\cdot 0$)	1·22	1·13	1·52	1·09	1·22	1·10
Percentage 'good' E.D.T. values (Ideal: 100%)	100	100	100	100	100	96
Percentage overall variation (Ideal: 0%)	38	30	21	15	9	19

one of the factors deciding this business. I must make reservations for the range, for the interval. I am not absolutely sure that the interval of decay time should be zero to minus 10 and I know that other workers are trying now with zero to minus 5, but I have one objection, the lesser the interval, the more the fine details of this curve become important and I do not want too much fine detailing. I want a stable figure which can be measured and compared with the reverberation time.

7

Acoustic Modelling—Design Tool or Research Project?

ALEXANDER N. BURD

Sandy Brown Associates,
Conway Street,
London, England.

The invitation to take part in this present symposium came at a time at which I was becoming increasingly aware of the problems of subjective appraisal in acoustic modelling and the *cri de cœur* which is contained in the title of my paper was a measure of my uncertainty as to the future. Of course the title was formulated a long time ago; the summary followed then the slides were drawn and finally the paper was written. Not a logical progression, but I suspect a fairly general one as the deadlines creep up upon you. You will be reassured to know that now I have found the answer to the question in the title of my paper and briefly it is 'yes' acoustic modelling can be a design tool and acoustic modelling can be a research project and in this paper I want to look at some of the factors which will govern the selection of the procedures that are used.

While I was thinking about what the subject of this paper should be, I looked back through some of the papers and reports on modelling that I had written during my time in the B.B.C. Research Department. It began some six years ago with acoustic models as an aid to studio design. This first paper was a consideration of the technique of modelling as it was described by Spandock and his workers and an assessment of the engineering feasibility of this type of modelling. We must have been adequately persuasive because the B.B.C. made the money available for us to start our work. At that time we stated quite clearly that the great advantage of modelling at a large scale, a scale of 1 to 8 in our case, and coupling this with a subjective appraisal, was the illumination of the need for deduction from objective parameters. By permitting people to hear the acoustic quality that would result, we would no longer be dependent on reverberation time calculations or visual evaluation of drawings; such naiveté, but I was younger then.

In 1969 came the famous non-reverberant music for acoustic studies—the first necessary step on the road towards subjective appraisal. I am well aware of the musical and to some extent the acoustical weaknesses of this recording but it was an improvement over the other sources of programmes that were available at that time and while it may never be number one in the charts, at least within our specialised circle, the sound of Mozart's Jupiter symphony has echoed or perhaps I should say fallen with a non-reverberant thump around the world.

During 1970 the requirements for all the various transducers, tape recorders, air-dryers and other hardware were considered and in general the requirements were satisfied.

In 1971 an $\frac{1}{8}$ scale model reverberation room was developed through Marks I, II and III by my colleague, Neil Spring and in 1972 we published an evaluation of the proving experiment. We took an existing B.B.C. orchestral studio and we compared it with its model and we showed that a recognisable characteristic of the acoustic quality of the studio was reproduced. In the next stage of the work we examined possible changes to the model. I do not intend in any way to refer to this in detail, but merely to say this studio was one in which we hung a lot of small diffusing reflectors above the orchestral position and another one in which we used a major orchestral reflector. A whole variety of possible modifications were examined which were expected to change the acoustic quality, and recordings of our music were made in each condition and then came the traumatic moment. In the proving experiment it was fairly simple, we were looking for a coincidence between the acoustic quality of two different conditions. In this case, where we had a variety of different changes, we were asking people to make an assessment of this change. Was it an improvement or a degradation? The simplistic view that you could assess the overall quality and reach a consensus view proved to be ill-founded. The average results for a group of observers for each of the changes that we examined was not significantly different from zero on a subject assessment scale which ran from better to worse, even though individuals had scored over a fairly wide range. The clue to this lay in the comments that people added on their sheets when they carried out the subjective assessment. Of course as we had modified the model we had also changed several characteristics of the acoustic quality and some of our subjects reacted in one way, some in another. One person would praise the added low frequency reverberance that we had given it, but considered that we had ruined the string quality and had led to an overall degradation. Another person who did not feel so strongly about the string quality was influenced by a completely different characteristic and so we got yet a different score from him. Inevitably at this stage we started on the slippery slope of subjective studies in which we examined a multiplicity of different acoustic qualities. More and more powerful statistical techniques were necessary to try and extract the latent meaning from the results and we finally reached the state which was described in the paper by Mr. Sproson and myself given at the last congress 'Analysis of factors relating to acoustic quality derived from recordings made in a model music studio.' Our design tool had become a research project.

TABLE 1

USER

ACOUSTIC CONSULTANTS		RESEARCH WORKS	
AUDITORIUM DESIGN	BROADCASTING STUDIO DESIGN	OBJECTIVE PARAMETER	SUBJECTIVE EVALUATION
REAL LISTENER	MICROPHONE	EXPERIMENTAL MEASUREMENTS	SUBJECTIVE STUDIES

Enough then of the background. Let us see if we can find out what are the requirements and the limitations of acoustic modelling as a design tool or as a research project. It is a bit of a truism to say that in the first instance it probably depends on who you are and what are your specialised interests. I have divided people into two categories, as shown in Table 1, and differentiated between those who may be more interested in a real auditorium in which people are going to listen directly and the broadcasting studio design consultant who is interested in a space which will be examined with a microphone. The difference between these two may very well be disappearing—the advent of artificial head recordings and perhaps some of the more modern ambio-sonic techniques may well mean that what one hears in the studio will be just as critical to the variations between side reflections and overhead reflections which is probably not the case in some of the studios at present. On the right hand side we have our research workers who will be interested in examining the model first to determine objective parameters by experimental measurements and perhaps subjective studies to look at subjective appraisal. The line down the middle is perhaps drawn even more truly than he realised by my draughtsman; of course these are not watertight compartments and I have no intention of trying to stop the acoustic consultant from doing research nor the research worker from designing auditoria.

Before we go on to consider and perhaps codify the model techniques available I ought to start by clearing the ground a little and defining what I am not going to talk about. I hope this will not produce too great a sense of dismay amongst any who find that their own work has been too easily

TABLE 2

CLASSIFICATION OF MODELS

SIMPLE	MATHEMATICAL	PHYSICAL		FULL SIZE
		SMALL SCALE	LARGE SCALE	
TWO DIMENSIONAL	COMPUTER ANALYSIS	1:20 or smaller	1:10 or larger	1:1

dismissed. At an International Symposium on Architectural Acoustics I am sure that you will forgive me for disregarding the use of models in subjects such as traffic noise propagation in urban areas, the outer ear or the underwater acoustic channel, to name only three from the last Congress. So the answer to our question will depend even more importantly on who you are and what acoustic modelling means to you and I would certainly be the last to suggest the type of modelling that we carried out at the B.B.C. was in any way the most important.

In Table 2 and the tables that follow an attempt has been made to classify the main techniques of modelling that fall into the field of architectural acoustics. I do not claim that my list will be in any way exhaustive merely that I was exhausted by the time I had finished it. Re-examination of the tables as I wrote the paper has shown several inconsistencies and omissions for instance which I shall discuss shortly; not all the simple methods that are classed on the left are two dimensional and there is a nasty hole between my small models and my large models although I do not know of anybody that has fallen into it.

Let us reduce once more the field by looking in particular (Table 3) at the extreme cases from those two. We have simple models, visual examination of drawings and the use of a ripple tank which are definitely the two

TABLE 3

MODEL EXPLORATION TECHNIQUES

SIMPLE	FULL SIZE
DRAWINGS	TEST ROOMS IN
RIPPLE TANK	ACTUAL BUILDING
OPTICAL	
ULTRASONIC PHOTOGRAPHY	

dimensional methods that I had in mind in the first place. However, optical models in which the surfaces are made mirrors or dark absorbing surfaces and lights used to examine the sound fields inside or the use of schlieren photography to examine ultrasonic waves in models are definitely three dimensional uses. Measurements, such as these last two cases, can obviously give information on the gross distribution of energy within the model or even on the directional distribution at a particular point but I think that their use as design tools or research techniques is very limited these days. Full scale examination of parts of a design, either in the laboratory or in a field study, may be the only way of examining the interreaction between various parts of the building. The introduction of structure-borne flanking paths in a partitioned space due to a lightweight floor, say, can really only be studied by constructing a part of that design and to take another example, the efficiency of various methods of sealing a partition to a hollow window mullion can again only truly be resolved by constructing part of it, but to call these modelling may be stretching the point although I accept that full-scale is the only ultimate test. However, I would class the techniques in this table, in so far as they are used, as design tools. Professor Lord may not altogether agree with that since some of the offices of his acoustic department are in full size, replicas of offices which were built as part of a research project.

Table 4 indicates in more detail the remaining categories in my first classification. Under the heading of mathematical models I have placed, particularly, the use of computer analysis of drawings to follow the path of sound rays through multiple reflections. Physical models, in their two categories, are constructed as a scale replica of the design or of the real building if for some reason you are examining the building rather than the design. In small scale modelling, as in the optical model that I mentioned before, no attention is paid to the detailed acoustic characteristics of the

Table 4

MODEL ANALYSIS

MATHEMATICAL	PHYSICAL small scale	PHYSICAL large scale
COMPUTER CALCULATION OF MULTIPLE REFLECTION PATHS FROM DRAWINGS	SPARK SOURCE MICROPHONE DETECTION OPTICAL RAY TRACING PHOTOELECTRIC DETECTION SMOKE VISUALIZATION	SPARK SOURCE MUSIC REPRODUCTION (REVERBERATION FREE) MICROPHONE DETECTION

surfaces in the completed building. The surfaces are categorised simply as absorbent or reflective surfaces and made so. For large scale modelling, care has to be taken to reproduce the absorbing characteristics of a surface at its scale frequencies so that it is adequately reproduced in the model. Each of these models can be tested by means of a spark source and a microphone to determine the impulse response and, perhaps more importantly, the directional distribution of the impulses that comprise the impulse response. For each small scale optical ray tracing techniques have been used with photo electric detectors or smoke visualisation and for the large scale models we have of course the reproduction of reverberation free music picked up by a microphone for subsequent examination by listening to it.

In Fig. 1 I have tried to gather together these analysis techniques and indicate which can be obtained from each type of model. Mathematical study can certainly give early reflection patterns in the direction of arrival of energy over the first few milliseconds and directional impulse responses again for the first few reflections. Small scale models can certainly give all those in the first three forms of results and can allow one to derive at least some, if not all, of the objective parameters that have been developed over the years. Finally the use of large scale models can obviously allow us to derive all the factors but they may be inefficiently used to get those which can be equally well obtained from smaller simpler models.

Having determined the impulse response, the technique of subjective appreciation by simulating the resulting sound field is shown in Fig. 2. There we have non-reverberant music reproduced from a multiplicity of loudspeakers in a non-reverberant environment and the strengths and delays of the sound from each channel are adjusted to correspond to the arrival time and the strength of the early impulse measured from a model or computed from a mathematical analysis. In this way the early sound field can be simulated and, since this fairly rapidly merges into a statistically random arrival of reflections, the programme is also passed through an artificial reverberation system and fed through several more loud speakers so that you get the full simulation of the sound field.

I suppose it is inevitable that in a series of papers such as we have here there will be some overlapping between different authors who worked in-

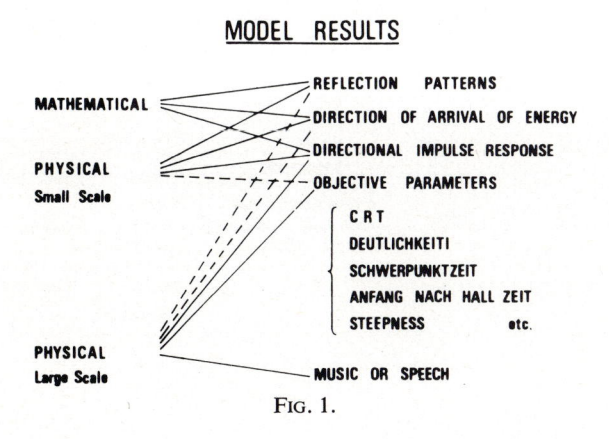

MODEL RESULTS

MATHEMATICAL — REFLECTION PATTERNS
— DIRECTION OF ARRIVAL OF ENERGY
— DIRECTIONAL IMPULSE RESPONSE

PHYSICAL
Small Scale — OBJECTIVE PARAMETERS

{ CRT
DEUTLICHKEITI
SCHWERPUNKTZEIT
ANFANG NACH HALL ZEIT
STEEPNESS etc. }

PHYSICAL
Large Scale — MUSIC OR SPEECH

Fig. 1.

SOUND FIELD SIMULATION

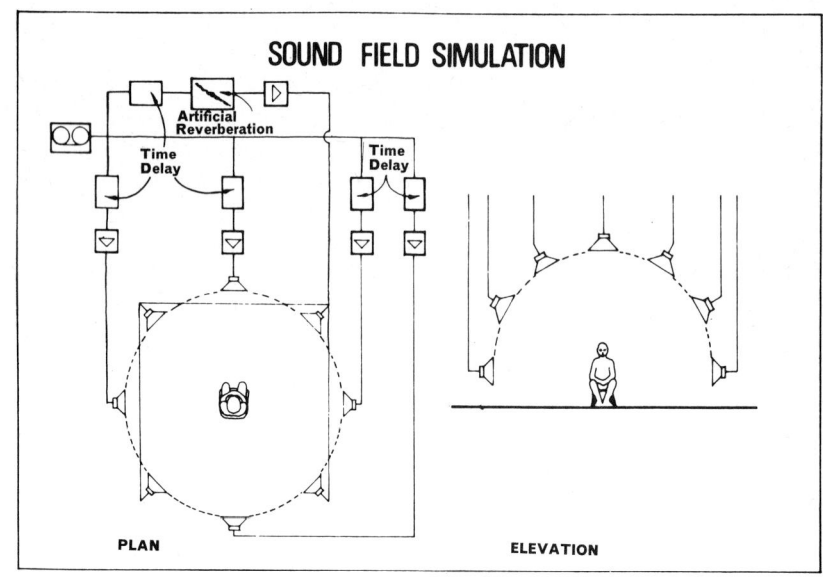

Fig. 2.

Table 5

SUBJECTIVE APPRAISAL

	REQUIREMENTS		
	source	model	hardware
COMPUTER SIMULATION	DIGITISED PROGRAMME	IMPULSE RESPONSE	COMPUTER, LARGE STORAGE CAPACITY
SOUND-FIELD SYNTHESIS	RECORDED PROGRAMME	DIRECTIONAL IMPULSE RESPONSE	DELAYS ATTENUATORS LOUDSPEAKER ARRAY
HEADPHONE LISTENING	REVERBERATION FREE PROGRAMME REPRODUCED FROM MODEL	MODEL ARTIFICIAL HEAD	HEADPHONES
LOUDSPEAKER LISTENING	DITTO	SPACED MICROPHONES	LOUDSPEAKERS

dependently on preparing the papers. I have tried to gather together in Table 5 the various techniques, the various ideas of subjective appraisal and the equipment requirements in each case. This includes the source of programme material digitised in the case of computer simulation or recorded in the other cases with reverberation free music, the requirements from the model for the information that you need to operate upon the source of sound in each of the first two cases, the computer simulation in the sound field synthesis, the impulse response or the directional impulse response and finally there are various hardware requirements. In the case of the computer simulation, obviously a large storage capacity computer, in the case of sound field synthesis, a large number of delays, attenuators and loudspeakers.

For the other two forms of listening appraisal, either with headphones or with loudspeakers, we require again the programme as a source in the model, and, if we are going to use headphone listening, then we require an approximation to the model artificial head. Obviously we are never going to get complete realism in this case because the head becomes far too small to get sufficiently sensitive microphones in. In the case of loudspeaker listening in the B.B.C. studies we have always used spaced omni-directional

TABLE 6

TIME SCALE FOR PHYSICAL MODELS			
Stage	Small Scale	Large Scale	
Drawings	1 week	4 weeks	4 weeks
Tendering	1 week	2 weeks	2 weeks
Materials Selection	—	1 weeks	4 to 8 weeks
Construction	3 weeks	8 to 20 weeks	12 to 24 weeks
Measurements	1 week	3 to 8 weeks	4 to 8 weeks
Evaluation and Report	1 week	2 weeks	4 to 10 weeks
Modification	2 to 6 weeks	?	?

microphones because it accepted the realities of the situation that we could get an adequate recording from these.

Now let us look at some of the constraints that are imposed upon our choice of techniques if we are interested in a design tool. First there is the architect's programme which seldom leaves any obvious gaps into which one can insert a model study. The work stages for a building project are shown in Table 6 and it is within the time scale of the outlined proposals and the scheme design that one has to try and insert the model study. By the end of the outlined proposal stage, there may just be enough information available to consider some of the general requirements for an auditorium. The scheme design stage produces specific proposals which are certainly adequate to produce a small scale model while sufficient flexibility still exists for modifications resulting from the study to be incorporated in the design. However, it is still difficult to get adequate time within this period to carry out the model studies so they have to be done as quickly as possible. There is an interesting interaction at this stage, since the information that is required for the construction of the model may well push the architect to make decisions before he would normally have made them and on the other hand the architect must be willing to accept proposals to changes stemming from the model studies, otherwise there is a danger that the existence of the model may well have frozen the design at a particular stage.

The time scale for the models is suggested in Table 7. Small scale models obviously require a relatively short time for the preparation of drawings for the tendering by the man who is going to make it and for the actual

TABLE 7

ARCHITECTURAL SERVICES

Work Stages For Building Project

- A. INCEPTION
- B. FEASIBILITY STUDIES
- C. OUTLINE PROPOSALS ⎫ MODEL STUDY
- D. SCHEME DESIGN ⎭
- E. DETAIL DESIGN
- F. PRODUCTION INFORMATION
- G. BILLS OF QUANTITIES
- H. TENDER ACTION TO COMPLETION

construction since this falls within the competence of a normal architectural model maker. Again the time required for measurements, which will normally be spark analysis, measurement of impulse responses is comparatively short and one can get through the work to a report stage in about 6 or 7 weeks. Each subsequent modification may well take a comparable time although it obviously depends on how fundamental the modifications that are required are, but one would have to accept that a sufficient period would be available for these modifications to be incorporated in the design.

When we look at the large scale models, I have divided them loosely into two categories. I am grateful to Stuart Harris for the title 'half tone' for the first category in which we take some account of the absorption characteristics of the surfaces but not complete detailed consideration and as for the other category, I shall call it the technicolour model, because we are going to reproduce, as far as we can, the entire absorbing characteristics. The small scale models incidently have become known as black and white models since they make no concessions to the detailed absorption coefficients.

So the time for drawings is very much greater here. We need a lot of information, a lot of interaction with the architects and the man who is going to make it has to take considerably longer to consider how he is going to do it and how much it is going to cost. The material selection in the case of a half-tone model is obviously comparatively limited and we only need to consider a limited number of materials, but in the case of a technicolour model there may be a lot of surfaces which have to be modelled. There may well be a lot of effort to be put into selecting the materials and into the construction which of course depends on the detail of the hall. Depending on the type of materials one is using it can take, in our experience, up to 24 weeks to produce a relatively half-tone model although it could be as short as 12. For a 1 to 8 scale model with all the absorbing surfaces included I would say that 6 months is not an unreasonable time.

And now to a very tricky question, 'How much is it all going to cost?' and my guesses become even more vague at this stage. For a small scale model, I would say a minimum must be about a thousand pounds on the basis of the effort and time that has to go in the requirements for draughtsmen, consultants and construction and could well be up to 5 times that sum. For the larger scale models, I would have thought that five thousand pounds was an absolute minimum and that twenty-five thousand would be not an unrealistic sum and they should well go substantially over that although obviously in this case you are getting much more into the field of a research project where large sums of money and considerable periods of time will be available to work with it.

Obviously I cannot tell you the answer to your particular requirements, that is a complex interaction between the time that is available, the money that is available and the schedules if it is an architectural design. All I have

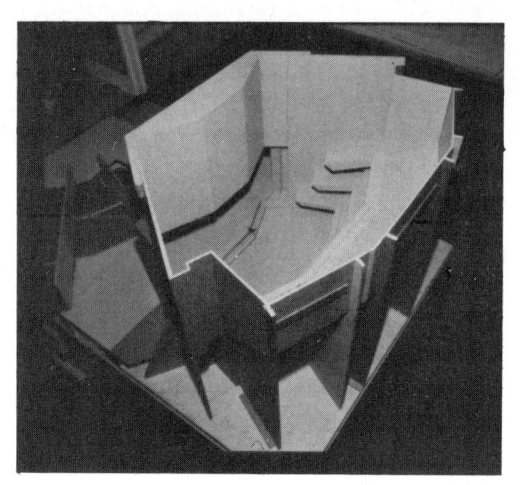

Fig. 3.

Fig. 4.

tried to do in this paper is to indicate where the choices lie and what may be the implications of your choice. For the Castle Terrace Theatre, Sandy Brown Associates have chosen to make a 1/50 scale model in which to study particularly the lateral reflection. Figure 3 gives an indication of the model. In fact I am glad to say that the architects have found it an extremely useful model too and in Fig. 4 which is shown by the courtesy of Robert Matthew, Johnson-Marshall, they show the use of this same model to give an architectural visualisation of what the inside of the space will be like.

We are proceeding at the moment to a half-tone 1/8 scale model in which we will carry out a variety of prediction techniques and this model also is intended to be a multi-purpose model and will be used to examine the air distribution by the mechanical services system that is to be incorporated.

8

Acoustic Scale Modelling Materials

B. Day

*Department of Architecture, University of Bristol,
Bristol, England.*

There are four major problems associated with the practical realisation of acoustic scale modelling principles: the selection of transducers; the modelling of the absorptive properties of internal surfaces; the modelling of air absorption, and the observation of room-acoustic phenomena at reduced time scale. This paper is concerned with the two central problems.

One of the main factors in the study of a phenomenon which forces us to resort to physical models is the complexity of available analytical solutions. Another reason is the ability we have with such techniques to make subjective studies of simulated music or speech which by-passes the whole problem of the selection of objective criteria, or so one might think. However, so long as we are unsure of the physical attributes which correlate with the required subjective attributes we cannot be sure how precise the physical modelling process needs to be, nor in which areas we should concentrate our attention to improve the accuracy. For example, if we were to decide that directional information (that is where the reflected sound comes from in the reflection pattern) was much more important than intensity variations then it would obviously be much more important to have correct directional properties in transducers, loudspeakers and receivers, than in the absorption coefficients of the surfaces.

An additional reason why we need to know a lot more about subjective response to sound fields is so that we can trim down the model to make it as simple as possible. We need an acoustic analogy with the architect's ability to convey subjective visual experience on a very crude model, which he calls a sketch. Consequently we need, in parallel with the development of accurate objective modelling techniques, more knowledge of the subjective perceptual process so we know where we can trim and where we can cut corners. However, since we cannot wait for a complete psychophysical description of the auditorium experience, at the moment we must utilise fully the physical concepts which we have, to guide the construction of models. We must therefore make as accurate a representation as possible of our conceptual model, that is, our picture of the physical situation in physical terms.

If we neglect (and it is a big omission) the properties required of transducers, this comes down to modelling the physical dimensions of the space, with the boundary conditions of its surfaces correctly represented at scale frequencies. We must also model the properties of the medium through which sound travels. I have to admit to taking something of a liberty with the title of this lecture in including air as one of the materials which we must model; I shall say a little bit about air absorption in a moment.

The least sophisticated model of the sound field which we have involves considering sound energy as a 'gas' of phonons diffused throughout the enclosure which is subsequently absorbed by the surfaces in the room to varying degrees. In fact one of the early theoretical derivations of Sabine's equation was lifted entirely from the study of the diffusion of gas from a space being evacuated by a vacuum pump. To satisfy this conceptual model, the surfaces only have to have the same random incidence absorption coefficients as the materials which they are intended to represent.

At a slightly more sophisticated level we consider the sound to consist of packets of energy which travel in straight lines or rays between the bounding surfaces of the room. At each encounter the ray obeys the laws of reflection and has its energy reduced by the absorption coefficient. In this case we need to know not just the random incidence coefficient but also the variation of energy absorption coefficient with angle.

Finally the most realistic conceptual model which we have involves consideration of the properties of the sound wave at the surface in terms of pressure and particle velocity, and characterises the surface by its complex impedance. This is influential in determining the phase of the reflected wave in addition to its intensity.

Ideally then a model boundary surface should represent at scale frequency the impedance and the variation of impedance with angle of incidence of its prototype at normal frequency. In the normal course of events the only effect of having incorrect impedances (provided that the combined effect of the real and imaginary parts produces the right overall absorption co-efficient) is to modify the effective position of the boundary very slightly. At the worst an incorrect reactive impedance might move the effective boundary by half a wavelength, in or out of the surface. Nickson and Muncey,[1] have discussed the precision required in the modelling of the impedances in relation to the accuracy of the geometric modelling, and this is not at all high. However, there are some circumstances in which precision in modelling the impedance becomes very important. An example is the ground effect in external noise propagation or, its equivalent, the Bekesy effect in auditoria. Attenuation rates are very high for sound passing close to an absorbing surface and theoretically strongly dependent on impedance

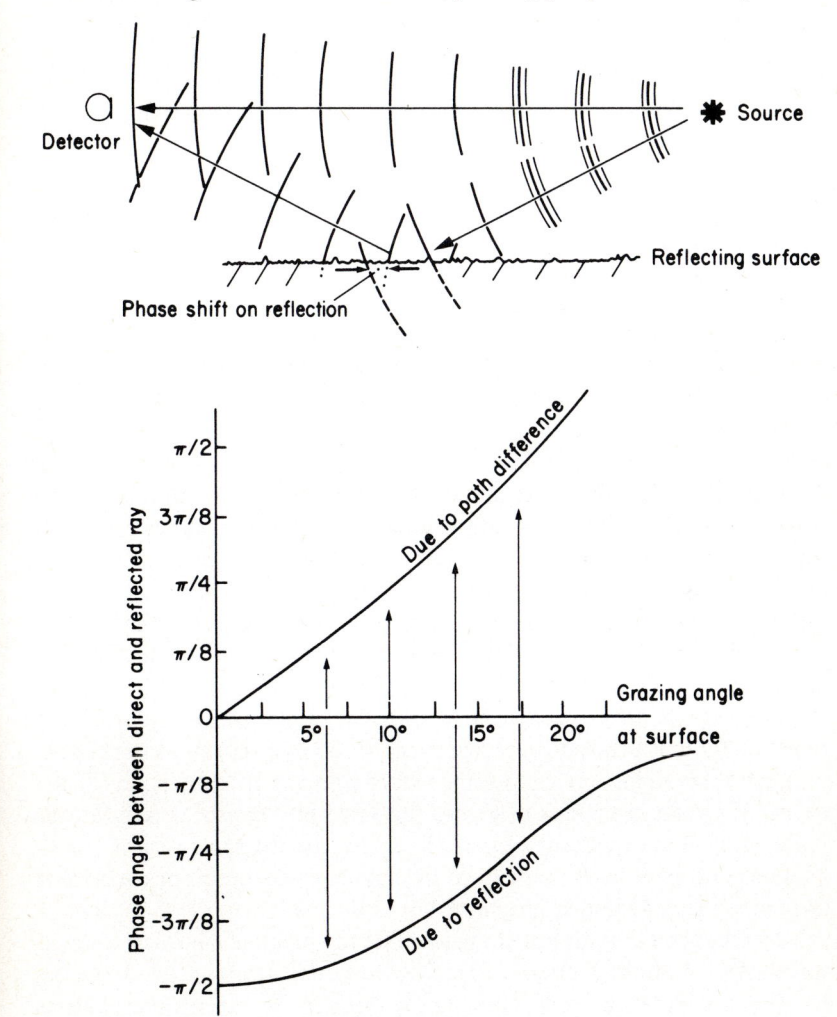

Fig. 1.

values. The origin of this effect is shown in Fig. 1. The upper figure represents a direct sound wave undergoing interference with a sound wave reflected from a nearby boundary. The phase angle between direct and reflected waves will depend on two factors: the path length difference and the phase-shift introduced by reflection at the boundary. As the distance between source and receiver increases, the angle of incidence increases and the magnitude of the path length difference gets smaller. The phase angle resulting from this difference steadily decreases, approaching zero for very large distances (grazing incidence). This is shown by the upper line in the lower figure. If the phase shift on reflection was zero, this would lead to a series of interference fluctuations, ending with constructive interference (reinforcement) at large distance. However, as the angle of incidence changes, the phase shift of the reflected ray varies, swinging from 0 degrees to 90 degrees at grazing incidence. This is represented by the lower curve of the lower figure; the shape of this curve will depend on the complex impedance of the reflecting surface. The resultant, total phase shift, which is represented by the arrows in the lower figure, will lead to destructive interference over a large distance, more or less independent of wavelength, and this is one explanation for the ground effect observed over soft ground—or over an audience, or the high attenuation rates sometimes observed in landscaped offices.

Now it is obvious that if you are modelling a situation of that sort you need to be very careful to get the impedances correct. Unfortunately techniques for measuring impedance variation with angle of incidence are only recently developed for prototype frequencies and not at all for model measurement. We have therefore got to fall back on measurements of random incidence absorption coefficient or normal incidence coefficient at lower frequencies, and combine this with a measure of self reassurance generated by looking at the mechanisms of absorption and telling ourselves that if we use the same mechanism, and get the same absorption coefficient, then hopefully it is being generated because the impedances are the same. And the same thing applies to some extent with variation in angle of incidence. We shall hope that if we use the same mechanism to achieve the same absorption coefficient then it is likely to have the same variation with angle of incidence as the prototype. In any case this need not concern us too much in modelling large spaces like auditoria.

Now I hope that this is not going to inflict too much of a burden on those who know all about it already or do not understand enough to start with, but I would like to go very quickly now through the various mechanisms of absorption and to see how these are likely to scale.

First of all—porous absorbers. If we have a porous absorber then there are a number of properties of the material that determine its impedance. We have the density and velocity of sound in air (ρ and c) which are invariant between the prototype and the model. We then have the porosity of the material σ, its flow resistance R, and a somewhat mystic quantity known as the structure factor (F) which we choose so as to get the right answer, rather than measure it. Then the layer has got some finite thickness l and the incident sound wave has some pulsatant or angular frequency ω.

$$Z = \frac{\rho c}{\sqrt{\sigma}} \cdot \sqrt{F} \cdot \underbrace{\left[1 - \frac{iR}{\omega \rho F}\right]}_{(a)} \cdot \coth \underbrace{\left\{\frac{i\omega\sqrt{\sigma F}}{c} \cdot \underbrace{\left[1 - \frac{iR}{\omega \rho F}\right]}_{(a)} \cdot l\right\}}_{(b)}$$

Now if we can keep the terms (a) (which appear in square brackets in the equation) constant in our scaling process, as well as the term (b) which appears in curly brackets in the expression, then we would hope that the impedance of the model material would match with the prototype at scale frequencies. To achieve this in a tenth scale model we should first remember that ω will be ten times larger. The density of the air (ρ) is the same in both model and prototype. The structure factor of the material (F) we hope will be the same if we have chosen a similar sort of material: fibrous mat, for example, to represent a fibrous mat. Consequently, to keep term (a) constant when ω increases by 10, R (which is the specific flow resistance of the material) must be 10 times larger. So to model a porous absorber at tenth scale one should select a material with 10 times the flow resistance of the prototype. If we now look at term (b) it will be seen that the thickness must be reduced to one tenth of that of the layer being modelled, for a tenth scale model, since again all the terms except ω are constant.

The absorption coefficient is given in terms of the real (u) and imaginary (v) parts of the impedance by the first expression in the next equation. The second expression deals with the variation of absorption coefficient with angle of incidence.

$$\alpha = 1 - \frac{(u-\rho c)^2 + v^2}{(u+\rho c)^2 + v^2}$$

$$\alpha = 1 - \frac{(u\cos\theta - \rho c)^2 + v^2}{(u\cos\theta + \rho c)^2 + v^2}$$

If the material is a porous material with a flexible frame then the situation is rather more complicated. Essentially in the theory of flexible frame porous absorbers the flow resistance term (a) is replaced by an expression containing a resistive coupling term g which has to do with the resistive drag between the flow of air in the pores of the material and the fibres of the material.

$$\sqrt{1 - \frac{iR}{\omega\rho F}} \quad \text{becomes} \quad \sqrt{\frac{1 - \frac{ig}{\omega\rho F}}{1 - \frac{ig}{\omega\rho_2 F}}}$$

The density of the fibrous material is represented by ρ_2. This is really getting a bit too complicated even to use for guidance in the selection of model materials. We just have to hope that if we select similar mechanisms we will discover a similar behaviour. As you will see in a moment the practical results do not justify carrying reliance on this principle very far. Now let us turn to the membrane type of absorber. In a recent paper, Ford and McCormick[2] have discussed the behaviour of membrane absorbers and it turns out that the impedance is a function of the terms G S N and T in the next equation which are not written out in full because they all involve, in summation form and in different linear combinations, the typical things you would expect from a mass spring system:

$$Z = u + iv = \frac{GS - NT}{S^2 + T^2} - i\left[\frac{NS + GT}{S^2 + T^2}\right]$$

Inertia: $\omega M A_{mn}$
ω = pulsatance, $M = d_1\rho_1$ = mass/unit area of membrane, A_{mn} = mode constant

Membrane stiffness: $\dfrac{B_{mn} D}{\omega a^4}$

$D = \dfrac{E d_1^3}{12(1-\sigma)}$ = bending stiffness, σ = Poisson's ratio, E = Young's modulus, a = side length, B_{mn} = mode constant

Cavity stiffness: $\dfrac{\chi P_0}{\omega d_2}$

d_2 = cavity thickness

First there is an inertia term representing the inertia of the membrane itself:

this involves the pulsatance again; M is the mass per unit area of the membrane and A_{mn} is a factor depending on the mode shape. Next there is a term for the stiffness of the membrane; the coefficient B_{mn} just depends on the mode in which the membrane is vibrating. D is the stiffness of the membrane, which depends on the modulus and the membrane thickness, and Poisson's ratio. The term in the fourth power of a is actually for a square membrane but similar terms can be written for a rectangular membrane, in terms of the dimensions of the side length. Finally there is a term for the cavity stiffness, that is, the compressibility of the air trapped in the cavity.

When we come to model such a system we must try to scale each of these terms separately. That is, in the scaling process we have got to make sure that each of the terms retains its value. For a tenth scale model we are going to operate at ten times the normal frequencies, so that ω will increase by a factor of 10. To ensure that the inertia term remains unchanged M must be reduced by 10. This will be achieved if we keep the same density of material by retaining the same material for the membrane (that is to say, for a wood membrane we use wood or for a steel membrane we use steel), but we decrease its thickness by a tenth; in other words we should adopt geometrical scaling. Furthermore, if we retain the prototype material in the model the Young's Modulus will be the same and the Poisson's ratio will be the same. By scaling the membrane thickness d_1 the bending stiffness goes down by a thousandfold (for tenth scale). The boundary lengths of the membrane are reduced by a factor of 10 and the pulsatance increased by a factor of 10. Consequently the membrane stiffness term also remains constant for geometrical scaling. The same is true of the cavity stiffness term, since γ and P_0 are constants; one should reduce the cavity depth to $\frac{1}{n}$-th for an nth scale model. Once again geometrical scaling looks as if it works.

Unfortunately, as we have described it, this system does not have any loss term at all. In practice there will of course be hysteresis in bending the membrane and elastic absorption at the edges. In order to cope with this in the equations we simply make Young's Modulus complex. The imaginary part then represents the damping. Unfortunately, although the real component may do so, the imaginary component is unlikely to remain constant when we scale the frequency. There will generally be an increase in the relative size of the loss term, so a simple geometrically scaled model of a membrane absorber using identical materials will be quite likely to have a broader width and a lower peak absorption than the prototype it is intended to represent. However, we can tune the model by including porous absorbers in the cavity or by not including those that may have been there in the prototype. So once again geometrical scaling looks as if it might work but we have got to modify the resistive terms in the system in order to obtain precise value.

Finally I will refer to the Helmholtz resonator system. Any Helmholtz resonator consists of a throat which has an effective length l', an area S while the volume at the back of the throat is V. There is a flow resistance in the throat (R_{throat}) which is largely responsible for the losses in the absorber. In a case where many resonators are distributed over a wall, A is the area of the wall associated with each resonator. The impedance of the wall surface is then given by the next expression and consists of 4 terms.

$$Z = A\left\{\left(\frac{\rho_0 \omega^2}{2\pi c} + R_{\text{throat}}\right) + i\left(\frac{\omega \rho_0 l^1}{S} - \frac{\rho_0 c^2}{\omega V}\right)\right\}$$

$$\quad\quad\quad (1)\quad\quad (2)\quad\quad\quad (3)\quad\quad (4)$$

If we adopt geometric scaling again, in which all the sizes are reduced by N then the volume, for example, goes down by N^3, the pulsatance has gone up by N so the fourth term as a whole is increased by N^2. This happens with terms (1), (3) and (4) so that they are all increased by N^2. However, A, the area per absorber decreased by N^2, so the thing as a whole (excluding the flow resistance in the throat) scales perfectly. Once again geometrical scaling wins the day except for the resistance term. We would wish the flow resistance to be increased by N^2 but it is unlikely to scale accurately because of changes in Reynold's number. So again we have to fall back on adjusting the resistive component empirically to get the absorption to scale correctly, although the frequency of maximum absorption should look after itself.

That concludes this brief survey of mechanisms available for absorption. Unfortunately, in no case does the analysis give us sufficiently clear guidance as to the selection of materials. Consequently, everybody that has worked in the field of scale models has eventually fallen back on to the exercise of building a library of modelling materials. One takes a potential modelling material and measures its absorption characteristics at scale frequencies and then looks for a matching prototype material. As each laboratory seems to use a different scale factor libraries are not very inter-changeable. However, we have recently completed a fairly extensive study of a range of materials at Bristol and I would be very glad to hand on to anybody that thought they could be of use to them. The figures we have obtained are for a variety of cardboards, hardboards, foamed plastics, cork felt, and various sorts of fabric. We have also examined the effect of perforated sheet covers and combinations of backing air spaces with these materials. The relevant physical properties have also been determined as far as possible.

Now a word about methods of measurement. We used a reverberation tank made of $\frac{1}{4}$ in steel plate, just to form a reverberant enclosure. One simply measures reverberation time in the appropriate frequency bands, with and without the sample. The I.S.O. procedure can be adopted with appropriate scaling of frequency. In our earlier tests we made a model of an existing full-size reverberation room, so that we could make direct comparisons between the absorption coefficients measured in the model and in the full-sized room.

Measuring reverberation time at low frequencies is fairly easy, since it is of the order of 1s or so (10s equivalent in a 1/10th scale model). One can simply use the normal procedure: noise generator, level recorder and so on in real time. When one moves to higher frequencies one can either go to fast writing speed instruments, logarithmic amplifiers and storage oscilloscope, or alternatively one can record the response of the room on a multi-speed tape recorder and then replay this with time scaling (lower speed) in order to bring all the frequencies and times into the normal (prototype) range. This is the technique that we have adopted because it seems to be quite useful for purposes of subjective testing as well.

The disadvantage of a 1/10th scale factor which arises with time scaling is that if you take the centre frequencies of the octave bands normally used at high frequencies and divide by ten, the octave bands which result do not correspond with I.S.O. preferred frequencies. This does not happen if you employ 1/8th scale however, or if one works with 1/3rd octaves. Another problem arises in trying to keep track of where you are in the test sequence when replaying a tape at low speed. We have developed a number of methods of overcoming this problem. In the latest, we pre-record all the test-signals at prototype frequencies on a slow moving tape, along with a commentary track which tells one which frequency band is being radiated and indicates the imminence of the cut-off of the signal. This tape, replayed at the scale speed, provides the noise signals which excite the reverberation room. At the same time the commentary is transferred directly to one track of a second time-scaling tape recorder, which also records the response of the model room. When you finally replay this recording at low speed, the commentary originally recorded with the test signals indicates where you are in the test sequence and what particular frequencies are being radiated and so on.

As an aid in the selection of materials for scale models we also thought that we should look at the usefulness of flow resistance measurements of porous materials. To this end we measured the flow resistances of some of the materials for which we had determined absorption coefficients. The measurements involved a very simple technique in which the sample was fixed across the mouth of an inverted bell jar, from which water ran at a measured rate. The pressure drop between the atmosphere and just inside the sample was measured with a manometer. It is a rather crude method because the velocity of flow is quite large, the pressure head varies to some extent as the water runs out, and you need a fairly sensitive micro-manometer to measure the small pressure differences. However, on the whole we obtained fairly consistent results. In some cases the plot of flow rate versus pressure difference turned out to be non-linear so that we suspect that some sort of non-linear flow phenomenon was taking place.

The next stage was to compare the measured results of absorption coefficient with the absorption coefficients calculated on the basis of the equation discussed earlier. We got very poor agreement indeed. In almost all cases the theoretical values were well below the experimental values. We are still investigating the origin of this discrepancy, which seems to be much worse than that found with normal materials at normal frequencies. One factor appears to be the choice of the value of the sound speed (c) substituted in the equation. By postulating isothermal sound waves in the material, and using the isothermal sound speed instead of adiabatic sound speed the theoretical absorption is pushed up quite a lot at low frequencies. But we are still some way from getting the right answers.

When one comes to consider the testing of materials for the representation of reflective surfaces rather than absorptive ones, obviously the technique for measuring absorption coefficients has to change. If one placed a sample of brick wall in a normal reverberation chamber it would be very difficult to measure the change in reverberation time. In exactly the same way, if we try to look at the absorption coefficients of our model wall materials in an I.S.O. type of test, adding sample areas to a reverberation chamber, the change is too small to be accurately measured. So for testing materials to be used to represent reflective surfaces, we adopted the procedure of constructing a complete box of the material and measuring the reverberation time of the empty box. Application of Sabin's equation yields the total absorption power of the walls of the box and its contents. By subtracting the absorption power of the microphones and loudspeakers and assuming that the walls all have the same absorption coefficient, its value may be calculated.

Our measurements for 9mm plywood showed that an unpainted surface gives a porous absorber type of curve, except at low frequencies where a rise in absorption corresponds well with the coincidence frequency for this material. Painting the surface brings down the absorption coefficient to about 0·03 at high frequencies: a single coat of paint is adequate, subsequent coats having little further effect. This gives quite a good simulation of wall surfaces.

Another problem is the modelling of separate objects in an auditorium, which we tackled some time ago now in connection with the absorption of a model audience. Here one meets again the problem of impedance representation because although individual auditors are sufficiently far from the walls not to exhibit resonant effects, as soon as you start to put auditors together the gaps between them then become comparable with a wavelength and so one needs to consider the impedance at the surface of an auditor produced by his clothing. We adopted a trial and error procedure to produce the final design: the successful pattern was Mk.10! It took a long search to achieve the required characteristics and we were very relieved to find that the models showed the same sort of variation of absorption with seating density as one observes with real auditors. This suggests that the impedances are about right. This seems to bear out the empirical rationale that if the absorption coefficients are right and the absorption is produced by the same mechanism, then it is likely that the impedances will be correct.

I would like finally to say a word or two about the air absorption problem. As a sound wave travels through the air it is attenuated exponentially at a rate determined by m in the equation:

$$I = I_0 \, e^{-mx}$$

where x is the distance the wave has travelled. In an nth scale model, m should be n times larger; in general it is much larger than this, since it increases very rapidly with frequency. Furthermore, the exponent m appears in the corrected Sabine equation as a term, $4mV$, where V is the volume of the room so that the air absorption error increases with the size of the room being tested.

Now the absorption of sound in air is due to two basic mechanisms known as classical absorption and molecular absorption. Classical absorption arises from the viscosity and heat conductivity effects in air, and there is nothing that can be done in modelling to control it. It is dependent upon

f^2 so that the absorption in a 1/10th scale model goes up by 100, whilst it should only increase 10 fold. Inevitably the absorption goes up by more than one would wish. Fortunately this is a much smaller component of the total air absorption than the molecular absorption, which is aided by the water vapour present in normal atmospheric air. The process is in fact one involving the conversion of vibrational energy in the oxygen molecules into translational energy, catalysed by the presence of water vapour molecules. At constant composition m is again proportional to f^2 but if one removes the oxygen the effect stops altogether or if one removes most of the water vapour it is slowed down so much that it is no longer significant. As it is a relaxation phenomenon, the variation of molecular absorption (m_m) with the frequency (f) obeys an equation of the form:

$$m_m = \frac{1 \cdot 25 \times 10^{-5} \times f_{max}}{1 + [f_{max}/f]^2}$$

Here f_{max} is the 'Napier' frequency, which is the frequency for which the existing humidity produces a maximum in the m versus humidity curve. There has been a good deal of discussion concerning the relation between the Napier frequency and the humidity, which has been summarised in a paper published a few years ago by Monk[3]. He proposes the following equation relating f_{max} to the humidity (h):

$$f_{max} = 175\,h + 6140\,h \left[\frac{1 \cdot 12 + h}{10 \cdot 4 + h} \right]$$

which may then be used to calculate the molecular absorption. We have produced tables of molecular absorption coefficient and classical absorption coefficient from these expressions which fit reasonably well with the practical measurement which have been reported[4].

The difference in the attenuation between the true-to-scale value and that which actually occurs in the model with normal air, can be tackled in two ways. One can attempt the difficult task of drying out the model, so as to reduce the humidity to a value where m is small enough. Another approach, which was described at the recent International Congress on Acoustics is to purge the model with nitrogen so as to reduce the oxygen concentration, which also reduces m.

Alternatively, if one is simply interested in objective measurements, it is possible by a simple piece of algebra to obtain an expression to correct for the excess absorption. The equation:

$$\frac{1}{T_p} = \frac{1}{nT_m} - 25 \left[\frac{m_m}{n} - m_p \right]$$

relates the reverberation time measured in the model (T_m) to the corresponding value in the prototype hall (T_p), where m_m and m_p are the values of the total air absorption at scale and prototype frequency, and n is the scale factor.

That concludes this review of absorption in acoustic scale models. There are some copies of the tables of air absorption and absorption coefficients available.

REFERENCES

1. Nickson, A.F.B. and Muncey, R.W., (1956). 'Accuracy of matching for bounding surfaces of an acoustic model', *Acustica*, **6,** 35–39.
2. Ford, R.D. and McCormick, M.A., (1969). 'Panel sound absorbers', *J. Sound Vib.*, **10,** 411–423.
3. Monk, R.G., (1969). 'Thermal relaxation in humid air,' *J. Acoust. Soc. Amer.*, **46,** 580–586.
4. Winkler, H., (1964). *Hochfrequenztech. u. Elektroakust.*, **73,** 121–131.

9

The Reverberation Process as Markoff Chain—Theory and Initial Model Experiments

R. GERLACH

*III. Physikalisches Institut
der Universität, Göttingen,
Federal Republic of Germany**

0. CRITICISM OF THE CLASSICAL REVERBERATION THEORY

The classical reverberation time formulas (Sabine, Eyring, etc.) have become, in recent years, increasingly criticised[1-4]. This has been the result of refined measurement techniques as well as the result of practical experience in the field. It has been found that the reverberation times of many modern concert halls have often turned out shorter than expected, an early example being the Royal Festival Hall in London.

In former times the discrepancies in measurements were believed to be a result of the difficulty in adequately taking into account all the absorbing elements. But now, not least due to computer simulation it is increasingly believed that the classical reverberation formulas incorporate fundamental mistakes. What is wrong with the formulae of Sabine and Eyring?†

In the first instance there is the problem of the mean absorption coefficient $\bar{\alpha}$, which is calculated as an arithmetic average. Such a calculation totally neglects the absorber location and is correct only if the sound field in the

---

* During 1973–4 Leverhulme Research Fellow in the Department of Building at the Heriot-Watt University.

† The formulae of Sabine and Eyring:

$$T_s = -\ln 10^{-6} \frac{1}{\bar{\nu}} \frac{1}{\bar{\alpha}}, \quad T_E = \ln 10^{-6} \frac{1}{\bar{\nu}} \frac{1}{\ln(1-\bar{\alpha})}$$

where $\bar{\nu} = \dfrac{cF}{4V}$ (c velocity of sound, F surface area, V volume of the room) is the mean collision frequency and $\bar{\alpha}$ is the mean absorption coefficient of the walls [5-7]. For the problematic mean collision frequency see, *e.g.* References [7-10].

room is completely diffuse, but normally such a sound field cannot be readily achieved. In fact an absorbing wall has a greater incident energy per unit area than a non, or less absorbing wall. This is easy to show by looking at a room with one absorbing and $n-1$ reflecting walls. Whereas the absorbing wall gets sound from the $n-1$ reflecting walls, a reflecting wall gets sound only from $n-2$ walls. The formulae of Sabine and Eyring assume a wall is hit by a sound ray with a probability proportional to its size and independent from the wall from which the sound ray is coming. This can hardly be correct, as (in an empty room) a sound ray cannot hit the same wall twice in succession.

If on the other hand we follow a suggestion from Atal and present the sound propagation in an enclosure as a Markoff chain, then the previous history of a sound ray is taken into account and so the room shape and the absorber location get influence on the reverberation time. The problematic averaging over the absorption coefficients of the individual surfaces is no longer necessary[11–14].

1. THE THEORY

1.1. The Reverberation Process as a Markoff Chain

There are several, mathematically equivalent, ways for presenting the reverberation process as a Markoff chain. The following method from the outset relates the physical description to the discrete mathematical description and thus we have the advantage of clarity.

Consider an enclosure bounded by n plain walls and excited by a sound source which is switched off at the time $t = 0$. We are interested in the progression of the total sound energy for $t > 0$. Using a geometrical-statistical method we assume the sound field can be represented by a sufficiently large number of sound rays or sound particles. In addition to the absorption coefficients (averaged over all directions) $\alpha_1, \ldots, \alpha_n$ of the n walls, let the transition probabilities p_{ik} be known, where a sound ray coming from the wall i has its reflection at the wall k.

As a homogeneous first order Markoff chain the reverberation process proceeds in discrete time intervals of length $\bar{t}$. Suppose the sound particles are distributed over the n walls. At times $\bar{t}$, $2\bar{t}$ and so on the distribution changes into a new one in accordance with the transition probabilities p_{ik}. The absorption of each wall is taken into account by multiplying before each transition the energy on the walls with the corresponding reflection coefficients $(1-\alpha_1)$.

The distribution of the initial energy $\varepsilon(t = 0)$ is given by the starting vector

The Reverberation Process as Markoff Chain

$$e^{(0)} = (e_1^{(0)}, \ldots, e_n^{(0)}) \tag{1}$$

where $e_i^{(0)}$ is the partial energy at the wall i. At the time $\bar{t}$ the first transition occurs and we obtain the distribution $e^{(1)}$, for which

$$e_k^{(1)} = \sum_{j=1}^{n} e_j^{(0)} (1-\alpha_j) p_{ik} \quad (k = 1, \ldots, n) \tag{2}$$

or as a matrix equation

$$\begin{array}{cccc} e^{(1)} & = & e^{(0)} & A & P \\ (\cdots) & & (\cdots) & \begin{pmatrix} \cdot & & \\ & \ddots & \\ & & \cdot \end{pmatrix} & \begin{pmatrix} \cdots \\ \cdots \\ \cdots \\ \cdots \end{pmatrix} \end{array} \tag{3}$$

(where $A = \mathrm{diag}\,((1-\alpha_k)_{k=1}^{n})$ is the diagonal matrix of the reflection coefficients and $P = ((p_{ik})_{i,k=1}^{n})$ is the matrix of the transition probabilities).

At the time $2\bar{t}$ the transition to the distribution $e^{(2)}$ occurs. This transition again follows the matrix AP

$$e^{(2)} = e^{(1)} AP = e^{(0)} (AP)^2 \tag{4}$$

Thus for $N\bar{t} < t < (N+1)\bar{t}$ (i.e. after N steps) we have the energy distribution

$$e^{(N)} = e^{(0)} (AP)^N \tag{5}$$

Obviously we get the total energy ε_N after N steps by adding the elements of the distribution vector

$$\varepsilon_N = \sum_{k=1}^{n} e_k^{(N)} \tag{6}$$

or with eqn. (5)

$$\varepsilon_N = \sum_{k=1}^{n} \sum_{i=1}^{n} e_i^{(0)} e_{ik}^{(N)} \tag{7}$$

where the $e_{ik}^{(N)}$ are the elements of the matrix $(AP)^N = E^N$. And for the total energy at time t we have

$$\varepsilon(t) = \varepsilon_N \bigg|_{N = [t/\bar{t}]} \tag{8}$$

(where [ ] indicates the nearest integer value). Thus it is with eqn. (7)

$$\varepsilon(t) = \sum_{k=1}^{n} \sum_{i=1}^{n} e_i^{(0)} e_{ik}^{(N)} \quad (N = [t/\bar{t}]) \tag{9}$$

In fact, there is a discrepancy with real conditions, in that the transition frequency $\bar{\nu} = 1/\bar{t}$ is the same for each sound ray. But, of course, this is a simplification, which is also incorporated in the Eyring formula. This can be proved, for example, by deriving the Eyring formula from a zero-order Markoff chain as a limiting case of this theory[14]. (In a zero order Markoff chain p_{ik} is independent of i.)

We also write the mean collision frequency in the usual form

$$v = \frac{cF}{4V} \tag{10}$$

(c is the velocity of sound, F the surface area, and V the volume of the room.)

The matrices of the transition probabilities are so-called non-negative matrices and in addition the sum of each row equals 1. Using the theory of the these matrices[15-17] it can be shown[14] that the reverberation process is nearly exponential and that the reverberation time is largely independent of the starting vector.

1.2. The Determination of the Transition Probabilities and the Reflection Law

The transition probabilities p_{ik} are calculated as being proportional to the solid angle spanned in average by wall k as viewed from the wall i. A computer programme was developed which calculates the matrix of the transition probabilities for an arbitrary convex room shaped by n plain walls. There also exists the possibility of computing the transition probabilities of a diffuse reflection in accordance with Lambert's cosine law, because it is often found that the structure of the sound field is more diffuse than purely geometrical[4]. Another computer program then calculates reverberation times or, by iteration, absorption coefficients.

In addition it is not very difficult to take into account the dependence of the absorption coefficient on the angle of incidence[14]. To do this we take the average of the absorption coefficient of the surface, not over the whole solid angle 2π, but only over the individual solid angles spanned as seen from the other wall. Each wall is then characterised by $n-1$ absorption coefficients, for which we are using the off-diagonal elements of the matrix A. However, for the present, the dependence of the absorption coefficients on the angle of incidence is not taken into account.

2. CALCULATIONS, COMPARISONS, MEASUREMENTS

2.1. Calculations, Comparisons

2.1.1. *Starting probabilities* Reverberation curves were calculated for several room shapes and absorber locations. These calculations confirmed the theoretical considerations that the starting probabilities have nearly no influence on the reverberation time[14]. Therefore the starting probabilities are calculated as being proportional to the area of the surfaces.

2.1.2. *Influence of absorber location* In this theory the reverberation time depends not on an *average* absorption coefficient but on the absorption coefficients of the *individual* walls. To study the influence of the absorber location consider three rectangular rooms each having one absorbing wall. If the mean absorption coefficient of the Sabine or Eyring formula is stated, then the reverberation times of these formulae are fixed. But each room still allows three different absorber distributions, that is, the small wall with a high absorption coefficient, the large wall with a low absorption coefficient, or the medium size wall with a medium absorption coefficient. So the theory based on the Markoff chain gives three reverberation times.

In Table 1 these reverberation times are compared with those of the Eyring formula. The reverberation times calculated with the transition probabilities show a marked influence of room shape and absorber location. The shortest reverberation times are consistently found if the absorber is on the smallest wall. This effect, measured for example by Kuhl[18], now can be formulated.

TABLE 1. Influence of the absorber location represented by 3 rectangular rooms (non-diffuse reflection)

Relative room dimensions	Absorbing wall	α	$T\ [1/\bar{v}]$	T/T_E
1·2:1·5:1	large	0·5	123·69	0·9433
	small	0·75	111·18	0·8479
	medium	0·6	117·16	0·8935
1·5:2:1	large	0·433	126·18	0·9623
	small	0·866	105·75	0·8065
	medium	0·65	111·28	0·8486
2:3:1	large	0·367	133·17	1·0156
	small	1·1	99·54	0·7591
	medium	0·733	105·91	0·8077

Constant in all cases $\bar{\alpha} = 0\cdot 1$
Sabine reverberation time T_S 138·16 $[1/\bar{v}]$
Eyring reverberation time $T_E = 131\cdot 13\ [1/\bar{v}]$
Reverberation time T by Markoff-Theory
α: absorption coefficient of the absorbing wall

The greatest difference relative to the Eyring formula is nearly 25% and is found for the room with the most unequal dimensions. With one exception only, all reverberation times are below Eyring's. This is in agreement with many field situations[19] and computer simulations made by Schroeder[3].

2.1.3. *Comparisons with computer simulations by Kuttruff* In Fig. 1 the Markoff theory is compared with computer simulations made by Kuttruff[4]. For some rectangular rooms Kuttruff simulated the sound propagation with about 10 000 sound rays using a digital computer. In each case one wall was totally absorbent, the other walls were assumed to reflect the sound rays in accordance with Lambert's cosine law. From the reverberation times obtained in this way 'effective' absorption coefficients were cal-

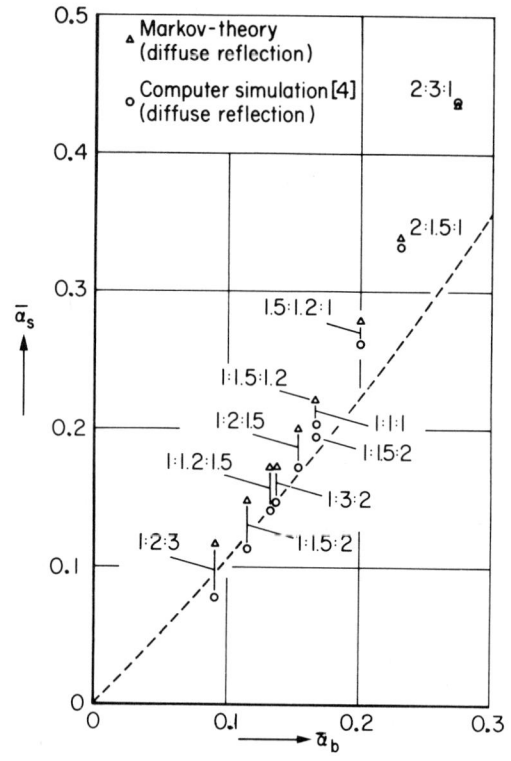

FIG. 1. Comparison between the reverberation theory based on a Markoff chain and computer simulations[4] represented by some rectangular rooms, given by their relative dimension, the first two numbers indicate the absorbing wall.
$\bar{\alpha}_s$ effective absorption coefficient calculated from the Sabine Formula.
$\bar{\alpha}_b$ absorption coefficient taken as arithmetic mean value.

culated employing the Sabine formula. In Fig. 1 these absorption coefficients $\bar{\alpha}_s$ are plotted as circles over the absorption coefficients $\bar{\alpha}_b$ calculated as the arithmetical average. The triangles indicate the effective absorption coefficients obtained in each case from the reverberation times calculated with the theory under the same conditions (*i.e.* diffuse reflection). The dashed line represents the Eyring formula.

In the first instance there are again marked differences between the theory and the classical formulas. The differences become greater for increasing $\bar{\alpha}_b$. Further, there is a very good correlation between the simulations and the Markoff theory for larger values of $\bar{\alpha}_b$. For smaller values the effective absorption coefficient of the simulations are below Eyring's. This unusual effect—with reverberation times longer than Eyring—could be a result of insufficient diffusion.

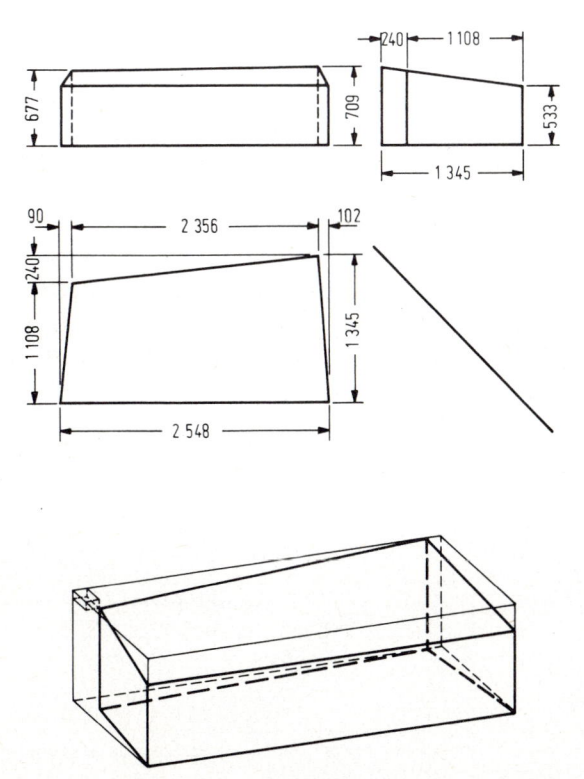

FIG. 2. The model reverberation room (dimensions in mm). The dimetric projection of the room is circumscribed with a rectangular prism to show the distortion of the room from a true rectangular prism.

2.2. Initial Model Experiments

2.2.1. *Preliminary Remarks* For the experimental proof of the theory model experiments have been chosen for two reasons. Firstly it was indicated (see section 2.1.3. and Fig. 1) that the classical formulas and this theory particularly contrast for greater values of $\bar{\alpha}_b$ and for rooms with more unequal dimensions. Consequently it is of interest to measure in a room of more extreme dimensions, for example 1:2:4, which was not available in full scale. Secondly model experiments have the advantage that the measurement conditions can be easily varied.

2.2.2. *The model room* In Fig. 2 we have a diagram of the model reverberation room. In order to avoid flutter echoes there exists no pair of parallel walls. The dimensional proportions of the room are about 1:2:4, the volume is 1·83 m^3. If this volume is compared with the volume of a concert hall then there results a linear scale factor of 25, and thus a measuring frequency of 25 kHz has been chosen.

The room is constructed of plastic coated chipboard. Thus the walls offer good reflection and are impervious to gas. Polyester foam of 6 mm thickness and 43 kg/m density was used as absorbing material. Measured by Brebeck[20], this material has an absorption of 100% at 25 kHz. However, experience has indicated that this is not valid for large angles of incidence.

2.2.3. *The measuring arrangement* The measuring arrangement is shown in Fig. 3. The room is excited in a corner by a spark source. The impulse response is filtered with a third octave filter and recorded on tape (Fig. 3a).

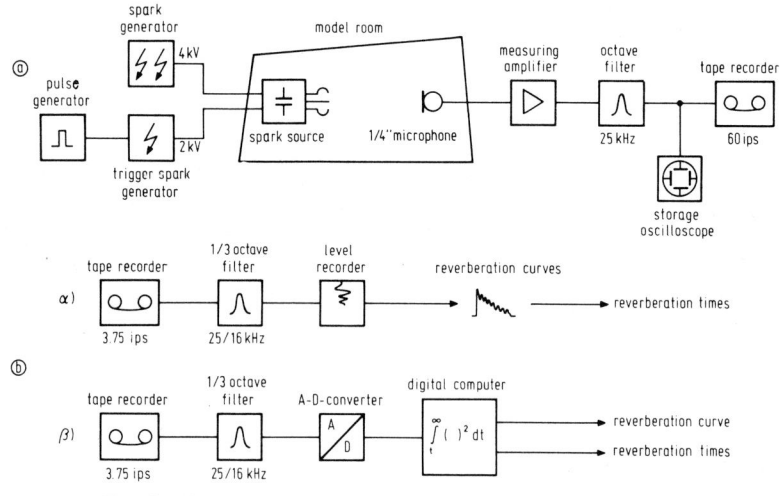

FIG. 3. The measuring arrangement and the evaluation method.

From the impulse response, played back at a lower speed, reverberation curves can be obtained in two ways: firstly in the conventional way with a level recorder (Fig. 3bα) or secondly—using a digital computer—with the method of the so called integrated impulse response[21] (Fig. 3bβ). Of course the second method is far more accurate.

2.2.4. *Steps to prevent the air absorption* At a measuring frequency of 25 kHz the air absorption is normally very high. This is a result of the oxygen and the water-vapour in the air. To avoid the air absorption these two parts of the air were removed by flushing the room with dry nitrogen from a steel cylinder.

Figure 4 indicates the dependence of the reverberation time τ of the absorptionless room on the volume $V(N_2)$ of the nitrogen streamed in. It appears that the curve becomes a constant at $V(N_2) = 25$ m^3. Therefore, the room is flushed with 9·5 m^3 nitrogen before the measurements are taken.

As a result of the nitrogen atmosphere the reverberation time of the empty room increases to 600 msec which is, in fact, 3 times longer than in air of relative humidity of 35% (see Fig. 5).

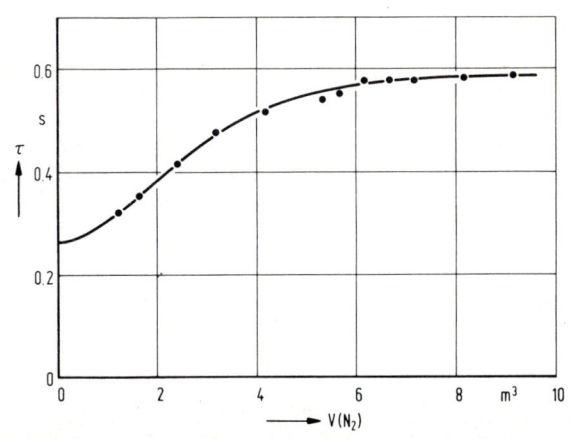

FIG. 4. Dependence of the reverberation time τ of the (nearly) absorptionless model room on the volume $V(N_2)$ of nitrogen streamed in (initially 50% relative humidity).

2.2.5. *Initial results and discussion* Measurements were carried out with one, two or three walls totally covered with the absorbing material. No attempt was made to increase the diffusion in the room. Reverberation curves obtained from these measurements are shown in Figs. 6, 7, 8. The reverberation curves obviously depend on the microphone position. This is naturally a disadvantage in verifying reverberation theories which describe the progression of the total energy.

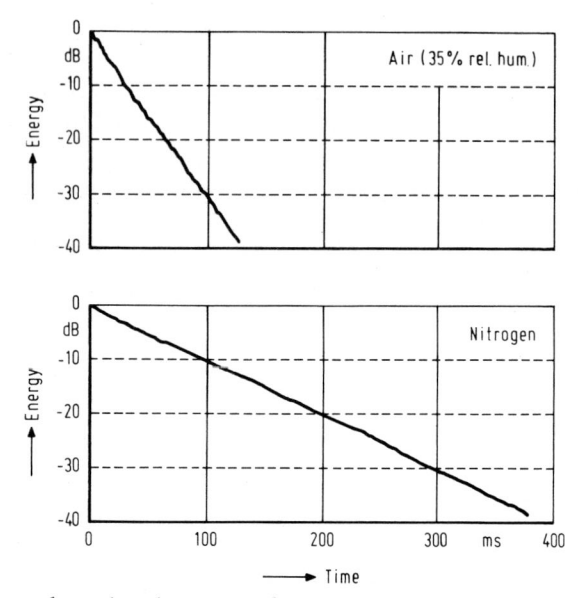

FIG. 5. Measured reverberation curves of the absorptionless model room (microphone at room centre).

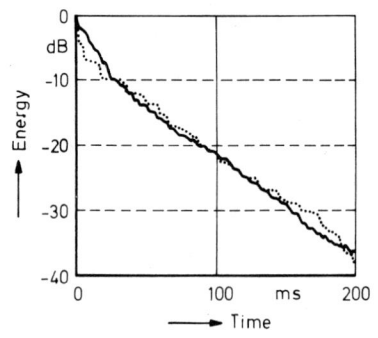

FIG. 6. Measured reverberation curves of the model room, absorber on wall 1.
Microphone position: —— room centre
· · · · room corner

In Fig. 9 the results of these initial measurements are compared with the calculations. The drawing shows reverberation times plotted over the mean absorption coefficient $\bar{\alpha}_b$. The measured reverberation times are obtained from the initial slope of the reverberation curves measured with the microphone at the centre of the room. (The initial slope is believed to be the most relevant part of the reverberation curve both for absorption measurements in the reverberation room[22] and for the subjective impression[23].) The numbers indicate the absorbing walls: wall 1 is one of the large walls (the floor in Fig. 2, relative dimensions 2 × 4), wall 2 is one of the medium

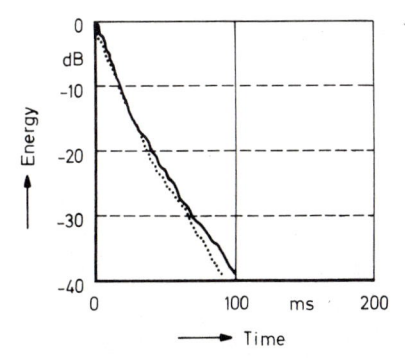

FIG. 7. Measured reverberation curves of the model room, absorber on wall 3.
Microphone position: —— room centre
· · · · room corner

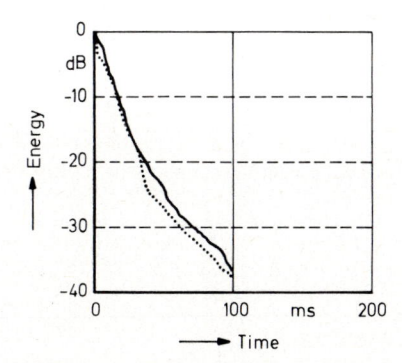

FIG. 8. Measured reverberation curves of the model room, absorber on wall 3 and wall 5.
Microphone position: —— room centre
· · · · room corner

size walls (the front wall in Fig. 2, relative dimensions 1 × 4), and wall 3 and 5 are the two small walls (the side walls in Fig. 5, relative dimensions 1 × 2). Measurements and theory correspond very well in some cases but in others, where only one wall is absorbent, they differ extremely. The shortest reverberation time is obtained if wall 3 is covered with the absorber and not if the larger walls 2 or 1 are covered with the absorber. The theory indicates this effect but is not capable of explaining its full extent.

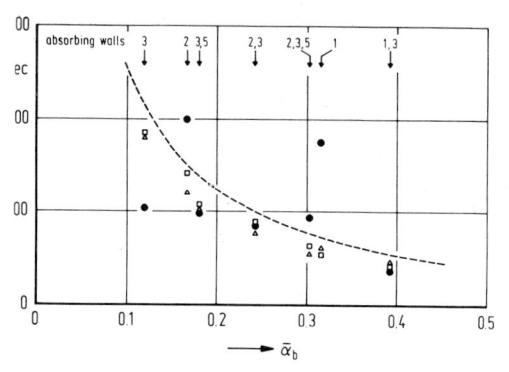

FIG. 9. Calculated and measured reverberation times τ of the model room for different absorber locations (indicated by the numbers) plotted over the mean absorption coefficient $\alpha_b = \varepsilon \bar{\alpha}_i F_i / F_i$.

△   calculated (diffuse reflection)
□   calculated (non-diffuse reflection)
●   measured (error smaller than symbol)
- - - Eyring formula

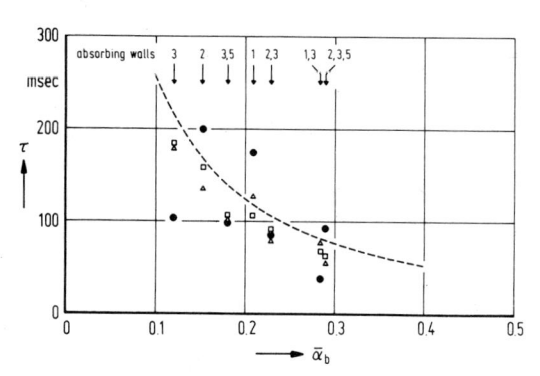

FIG. 10. As Fig. 9, but the dependence of the absorption coefficient on the angle of incidence is approximately taken into account.

If an approximation (details see Reference 14) of the dependence of the absorption coefficient on the angle of incidence is taken into account (see sections 1.2 and 2.2.2.) then the agreement between measurements and calculations is closer but still not conclusive (Fig. 10).

Obviously these initial measurements are still not capable of confirming the theory, due to the low degree of diffusion in the room. Therefore, the diffusion should be increased by stages. Using diffusing elements on the side walls the coupling between the sound field and the floor could be improved[24]. Besides this the theory will be applied to more complicated model rooms and full scale rooms.

3. CONCLUDING REMARKS

The model measurements reconfirmed that reverberation formulas are required which take into account room geometry absorber location and the degree of diffusion.

The presented theory is sensitive to room shape and absorber location. The theory needs less diffusion than the classical formulae because averaging over the individual absorption coefficients is not necessary. A certain degree of diffusion is, of course, an assumption for applying the theory because it describes the progression of the total energy in the room. But if it is possible to dispense with the simplification that the free path is the same for all sound rays then the required degree of diffusion could be further decreased.

REFERENCES

1. Gomperts, M.C., (1965–6). 'Do the Classical Reverberation Formulae still have a Right for Existence?' *Acustica*, **16**, 255–68. Kuttruff, M., (1965–6). Comments, *Acustica* **16**, 369–70. Gomperts, M.C., (1966). Reply to the comments, *Acustica*, **17**, 241–2.
2. Wagner, J., (1971). '*Mathematical and logical Criticism of the Classical Reverberation Theory*', Rozpravy Československé Akademie VĚD Rada Matematických a Přírodních VĚD, Ročnic 81 Sešit 3.
3. Schroeder, M.R., (1970). 'Digital Simulation of Sound Transmission in Reverberant Spaces.' *J. Acoust. Soc. Amer.*, **47**, 424–31.
4. Kuttruff, H., (1971). 'Simulierte Nachhallkurven in Rechteckräumen mit diffusem Schallfeld', *Acustica*, **25**, 333–42.
5. Meyer, E. and Neumann, E.G., (1967). *Physikalische und technische Akustik*, Vieweg, Braunschweig.
6. Cremer, L., (1961). *Statistische Raumakustik*, S. Hirzel, Stuttgart.
7. Kuttruff, H., (1973). *Room Acoustics*, Applied Science Publishers, London.
8. Kosten, C.W., (1960). 'The Mean Free Path in Room Acoustics,' *Acustica*, **10**, 245–50.

9. Kuttruff, H., (1970). 'Weglängenverteilung und Nachhaliverlauf in Räumen mit diffus reflektierenden Wänden,' *Acustica*, **23**, 238-9.
10. Kuttruff, H., (1970). 'Weglängenverteilung in Räumen mit schallzerstreuenden Elementen,' *Acustica*, **24**, 356-8.
11. Gerlach, R., (1973). 'Der Nachhallvorgang als Markoffsche Kette,' in *Fortschritte der Akustik*, (DAGA, Aachen, 1973). VDI-Verlag, Düsseldorf, pp. 427-30.
12. Schroeder, M.R. and Gerlach, R., (1973). 'Diffusion, Room Shape and Absorber Location—Influence on Reverberation Time,' 86th Meeting, Acoust. Soc. Amer., Invited Paper Q7.
13. Schroeder, M.R. and Gerlach, R., (1974). 'Diffusion, Room Shape and Absorber Location—Influence on Reverberation Time.' *J. Acoust. Soc. Amer.*, **56**, 1300.
14. Gerlach, R., Mellert, V. 'Der Nachhallvorgang als Markoffsche Kette— Theorie und erste experimentelle Überprüfung.' *Acustica*, to be published.
15. Zurmühl, R., (1964). *Matrizen* (4. Aufl.) Springer, Berlin.
16. Brauer, A., (1964). 'On the Characteristic Roots of non-negative Matrices,' in *Recent Advances in Matrix Theory*, H. Schneider Ed., University of Wisconsin Press, Madison.
17. Darroch, J.N. and Seneta, E., (1965). 'On Quasi-Stationary Distributions in Absorbing Discrete-Time Finite Markoff Chains.' *J. Appl. Prob.*, **2**, 88-100.
18. Kath, U. and Kuhl, W., (1961). 'Einfluss von Streufläche und Hallraumdimensionen auf den gemessenen Schallabsorptionsgrad,' *Acustica*, **11**, 50-64.
19. Schroeder, M.R., (1973). 'Computer Models for Concert Hall Acoustics.' *AJP*, **41**, 461-71.
20. Brebeck, D. 'Die Schall- und Ultraschallabsorption von Materialien in Theorie, und Praxis, insbesondere im Hinblick auf den Bau akustisch ähnlicher Modelle im Maß-stab 1:10.' Dissertation München, Feb. 1967 Verzeichnis der Hochschulschriften, U 67. 12 144.
21. Schroeder, M.R., (1965). 'New Method of Measuring Reverberation Time,' *J. Acoust. Soc. Amer.*, **37**, 409-12. Smith, P.W., 'Comments' and Schroeder, M.R., 'Response', (1965). *J. Acoust. Soc. Amer*, **38**, 359-61.
22. Kuttruff, H., Josefie, M.J. (1969). 'Messungen des Nachhallverlaufs in mehreren Räumen ausgeführt nach dem Verfahren der integrierten Impulsantwort.' *Acustica*, **21**, 1-9.
23. Schroeder, M.R., (1965). 'Subjective Reverberation Time and its Relation to Sound Decay.' 5th Congrès International D'Acoustique, Liège G 32.
24. Meyer, E., Kuttruff, H. and Lauterborn, W., (1967). 'Modellversuche zur Bestimmung des Schallabsorptionsgrades im Hallraum.' *Acustica*, **18**, 21-32.

10

The Influence upon Auditorium Reverberation Time Caused by Different Positioning of Absorption within the Fly Tower

PAUL NEWMAN

*Department of Architecture,
Heriot-Watt University,
Edinburgh, Scotland.*

I should like to take this opportunity of briefly describing the research in architectural acoustics currently being undertaken in the Department of Building at the Heriot-Watt University.

Whilst model scale acoustic research work has been pursued for many years now by my colleagues in the Department, this has taken the form of sound insulation studies rather than auditorium acoustics. However, in the last few years we have become increasingly involved in the latter field and we are at present engaged in model studies of the proposed Opera House to be built in Edinburgh (Castle Terrace Theatre Project).

Prior to the construction of the $\frac{1}{8}$th scale Opera House model at the University, we constructed another model of similar scale of an existing condition. The model is a simulation of Leith Town Hall, an auditorium used for concerts and chamber music, particularly during the Edinburgh Festival.

The Leith Town Hall model has been used for comparative measurements in the model and in the Hall. It has enabled us to develop measurement techniques which are now being applied with the Opera House model.

The studies to date in the Leith Town Hall model have been mainly concerned with the coupling effects between the fly tower and auditorium volumes of the Leith Town Hall model. Conflicting opinions have been expressed as to the extent of the effect a reverberant fly tower may have upon the decays measured in the auditorium. In predictions of reverberation time it has been the traditional approach to consider only the auditorium volume and to treat the proscenium opening as an absorbent boundary. Knudsen and Harris, Ingerslev and others have given absorption coefficients for such openings. However in our own studies an approach was made on the basis that the fly tower has a volume sometimes exceeding that of the adjoining auditorium. If there is a significant transfer of energy through the proscenium opening, the fly tower and the auditorium may more properly be treated as a single volume for the purposes of predicting reverberation times.

Suppose the fly tower contains a considerable amount of absorbent material such that the reverberation times within the fly tower when closed off from the auditorium may be similar to those in the isolated auditorium. With the proscenium open the volume of the auditorium could be increased several times by the addition of the fly tower and the surface area to volume ratio correspondingly reduced.

Preliminary studies (See Table 1) with the Leith Town Hall model have yielded tentative support to the idea that the fly tower and the auditorium

TABLE 1. *Comparative Reverberation Times (seconds) ($\frac{1}{8}$ full scale model: Leith Town Hall). Mean values for four microphone positions in the stalls area of the auditorium. Reverberation times have not been corrected for air absorption effects*

	Equivalent full scale octave band centre frequencies					
	125 Hz	250 Hz	500 Hz	1 KHz	2 KHz	4 KHz
Proscenium open, fly tower with no added absorption	2·95	2·4	2·0	1·8	1·15	1·05
Proscenium closed with timber panel	2·65	2·2	1·95	1·6	1·05	0·8
Proscenium open, fly tower highly damped	2·4	2·05	1·55	1·4	1·1	0·85

should more properly be treated as a single volume or at least as two strongly coupled sub-systems. It should be noted that the fly tower volume of the Leith Town Hall is small relative to the volume of the auditorium and is not at all typical of modern auditoria. In our current studies in the model of the opera house (where the volume of the fly tower is relatively much greater), the effect of the fly tower volume can be clearly seen in comparative measurements of reverberation time, even though the fly tower is highly damped. (See Table 2).

Since these measurements are comparatively recent a more detailed analysis cannot be presented at this stage. Nevertheless the release of data at such an early stage in the programme of tests can perhaps be justified by the significance of such an effect upon the basic design of auditoria. It is hoped that a full paper may be published next year on this research.

In addition to conventional reverberation time measurements we are carrying out steady state measurements and recording impulse responses. We have considered the application of Schroeder–Kuttruff techniques for measurement of early decay times and early energy fractions. However we decided not to use analogue instrumentation for this purpose.

TABLE 2. *Comparative Reverberation Times (seconds) ($\frac{1}{8}$ full scale model: Edinburgh Opera House). Mean values for four microphone positions in the stalls area of the auditorium. Reverberation times have not been corrected for air absorption effects*

	Equivalent full scale octave band centre frequencies					
	63 Hz	125 Hz	250 Hz	500 Hz	1 KHz	2 KHz
Proscenium open, fly tower highly damped	4·63	3·73	2·75	2·45	1·90	1·35
Proscenium closed with heavy fabric curtain	3·88	3·13	2·48	2·00	1·60	1·20

We considered that it would give us greater flexibility in the study of impulse response and decay processes if we were to use digital data processing. Responses to spark impulses are captured using a Bruel & Kjaer Digital Event Recorder. The captured response is inspected by taking the analogue output of the recorder to a storage oscilloscope. Once we are satisfied with the signal we then transfer it directly from the digital output of the recorder to an on-line Wang 2200 computer. This link between the two instruments affords considerable flexibility in the study of impulse responses and decays. A feature of the system is the ease with which the length of signal to be studied can be controlled. Course control of the timing of the signal within the memory of the digital event recorder is achieved with the after trigger recording control. Visual inspection of the signal then enables one to specify the beginning and end of the signal to be processed by the computer to a fraction of a millisecond.

The use of the dedicated computer has greatly extended the utility of our architectural acoustics laboratory. The video facility and software have been of great benefit in the development of programmes. We have found the main limitation to be the relatively slow transfer of data; it taking some minutes to transfer 10k samples from the digital event recorder to the Wang 2200.

At the present time the system is being used to produce squared and integrated early decay times and early energy fractions with graphical output of decay curves. The potential for other types of analysis is clear.

I should like to acknowledge the support given by my colleagues working on this project and also the Corporation of Edinburgh under whose contract this work on the Opera House has been made possible.

11

Vern Oliver Knudsen

A MEMORIAL TRIBUTE
R. W. B. Stephens

President of the Institute of Acoustics,
and Department of Physics,
Chelsea College,
University of London,
England.

I deem it a great privilege to give this memorial lecture in honour of a man who for many decades has been the acknowledged world leader in architectural acoustics. As an overall introduction to his achievements I cannot do better than quote from the citation for Vern Oliver Knudsen when he received the gold medal award of the American Acoustical Society in 1967. His qualifications for the award were 'For original research into the propagation of acoustical waves through air and through the sea; for his contributions to the understanding of the communication of speech and music and his expert application of this knowledge in the field of hearing and of architectural acoustics; for the dissemination of his research findings by superior teaching and authorship; and for his gentle but effective guidance during a lifetime of service to the Acoustical Society of America, as founder, officer and senior counsel'.

Vern Oliver Knudsen was born in Provo, Utah on the 27th of December 1893 and died on May 13th of this year from pneumonia, a complication of Hodgkin's disease from which he suffered during the last months of his life. He is survived by his wife, Florence and their children Morris, Robert and Margaret and two grandchildren. The time feature of his passing had a similarity to that of his close friend and colleague, Leo Delsasso who, you may remember, died on his journey to Budapest for the 7th I.C.A. meeting, at which he was to have contributed three papers.

Vern Knudsen entered Brigham Young University in 1911 with the idea of specialising in mathematics or engineering but he asserts himself it was the result of Professor Harvey Fletcher's influence that he finally gravitated to physics. Following his B.A. graduation in 1915 he served as a Mormon missionary and as acting head of the Northern State mission in Chicago. Later during World War I he became involved in the problem of speeding of cable telegraph transmission and to use his own words 'This was considered necessary because the existing cable was not adequate to carry the wordy communications between Woodrow Wilson and Lloyd George'.

During World War II he helped to organise what has become the Naval Undersea Research and Development Centre at San Diego and he

VERN OLIVER KNUDSEN (1893-1974). In front of the UCLA Physics Building which bears his name.

served as its first Director of Research. Together with Delsasso he guided much of the earlier investigations of the propagation of sonar signals and of the influence of ambient sound in the sea.

At the time when he joined Bell Telephone Corporation in 1918 there was considerable activity in the new developing vacuum-tube technology and Harvey Fletcher was using thermionic valves in his hearing studies at the Bell Laboratories. This experience was to be of immediate value to Knudsen when, in 1919, he became a graduate student at the University of Chicago. It is said that 'Wisdom is knowing what to do; skill is knowing how to do it; virtue lies in doing it and doing it well' and these you will learn were attributes of Knudsen. He had by now become most interested in acoustics as a career and so was disappointed when his supervisor, the great Professor R. A. Millikan, decreed that his doctoral dissertation should be 'On the contribution of electrons to the specific heat of metals'. Fortunately for the world of acoustics Millikan went abroad for three months and Professor Gale, the acting chairman of the physics department, was very considerate and knew that Knudsen, now married, wanted a project to which he could reasonably see an end in three years. Professor Gale encouraged Knudsen to take up a problem of his own choice and to advance the investigation so far that Millikan 'would not have the heart to disapprove' on his return. This benevolent action of Professor Gale incidentally had its impact on Knudsen's own attitude towards graduate education during his 24-year term (1934–58) as Dean of the Graduate Division at UCLA. Aided by the skills he acquired at Bell Laboratories and by dint of hard work he was able to make useful headway on the problem of sensibility of the ear to small differences of intensity and frequency. Millikan, on his return, was naturally surprised by the events which had taken place but allowed Knudsen to carry on with his acoustical work and, moreover, introduced him to Dr. G. E. Schamburgh one of the foremost otologists in the United States. Together Schamburgh and Knudsen carried out investigations on the sensibility of pathological ears to small differences in loudness and pitch, and on cases of diplacusis (which is the hearing of the same tone at a different pitch in each ear).

In 1922 Knudsen received his Ph.D. in physics *magna cum laude*, but to the surprise of everyone he refused offers from Bell Telephone Laboratory and from the University of Chicago and accepted the post of instructor at the newly formed University of California Southern Branch, now known as UCLA. The physics staff consisted of an associate professor, two assistant professors, an associate and Delsasso (who although only a sophomore student, acted as an assistant to the departmental chairman). This was the beginning of the lasting close personal and professional partnership between Knudsen and Delsasso.

Using the simple tools of an organ pipe and a stop-watch Knudsen immediately started his acoustic work by investigating the acoustical features of the many auditoria and halls in the Los Angeles area. He also linked with Dr. I. H. Jones, an otologist on studies of normal and impaired hearing, and this led to the development of the Knudsen-Jones audiometer. This was the first instrument that enabled the otologist to make a differential diagnosis between conductive (middle ear) and perceptive (cochlea) deficiency of hearing, and to test the cochlea directly. Twenty-six of these audiometers were constructed in Knudsen's backyard but were only made available to doctors who agreed to use them for research.

I have spoken generally about Vern Knudsen's career so I will now become a little more specific and comment firstly on his ability as a teacher. His courses on acoustics at the University were regarded as models of careful preparation and moreover covered all aspects of the subject: physical, physiological, psychological, architectural, *etc.* They were perhaps the only advanced undergraduate courses in the physics department where one could find students from a wide range of other disciplines. Knudsen's teaching was highlighted by his enthusiasm and the way he inspired his students. Among the list of former students who have attained prominence in the acoustic world may be mentioned Cyril Harris, Robert Watson, W. A. Munson, Norman Watson, Richard Bolt, Paul Veneklasen, Isadore Rudnick and Robert W. Leonard, whose tragic death in 1966 greatly affected Knudsen. The following extract from Knudsen's philosophy of graduate and higher aspects of education demands our consideration today. He says 'Although the scholar's search for truth is a worthy and laudable end in itself, and already this search has given us a great reserve of knowledge, there is a growing need for concern respecting the best uses of this knowledge. A centre of vigorous and creative intellectual activity is indispensable in a graduate school, but, more than ever in the past, we should strive to keep this activity geared to the good and the aesthetic, to that which will serve and uplift humanity'. Another extract worthy of mention says 'The assignment of appropriate and worthy problems for doctoral dissertations is one of the most serious and important functions confronting the graduate instructor. Frequently the entire career and social usefulness of first-rate scholars are determined by these assignments'.

Knudsen's main contribution to acoustic literature was entitled *Architectural Acoustics* and many would regard it as the 'bible' of the subject and I am proud to own a copy of the original edition. It is interesting to note that at the time of the publication, 1932, there had been quite a spate of acoustical publications, for 1931 saw the appearance of *Planning for Good Acoustics* by Hope Bagenal and Alexander Wood and in the same year a 40-page bibliography of over 400 references on Acoustics of Buildings by F. R. Watson appeared in the *Journal of the Acoustical Society of America*. In 1930 both Watson's *Acoustics of Buildings* and A. B. Wood's *Text Book of Sound* were published while in 1929 the classical volume written by Knudsen's mentor, Harvey Fletcher, entitled *Speech and Hearing*, made its appearance. About this time too, the Noise Abatement Commission, Dept. of Health, New York City produced a 300-page document on *City Noise* which was the result of a campaign against noise mounted in the 1920's, a campaign which Knudsen strongly supported. On his retirement in 1960 he was able to renew this activity and give strong support to efforts for the improvement of environmental quality.

Knudsen's other definitive book *Acoustical Designing in Architecture* published in 1950 and written jointly with one of his old students, Cyril Harris, became a standard work of reference on the subject. This also reminds us that Knudsen participated in the acoustical design of more than 500 structures and a few of these which he singled out particularly were:

University of Washington Music Buildings, (1946);
General Assembly, U.N. (with other consultants), (1948–50);
Schonberg Hall (Music Building) UCLA, (with Leonard and Delsasso), (1955);
Honolulu Concert Hall Theatre, (1964);
Atlanta Memorial Fine Arts Centre, (1968);
Phoenix Convention Centre and Concert Hall, (1972);
University of Akron, Ohio, Performing Arts Hall, (1973).

It was in 1929 that sound made its intrusion into the motion picture industry and Knudsen was consulted in the design of stages for sound recording and became involved with such well-known studios as M.G.M., Paramount Pictures, Fox, Universal, Warner Bros. *etc.* For this work it was important to have a fairly accurate knowledge of the acoustic absorption coefficient of materials used in construction and this led to the need for suitable measurement facilities. These were provided in the new buildings of UCLA, when it moved to Westwood, in the form of a double-walled reverberation chamber and two smaller adjacent measuring rooms also double-walled. This development leads to the point of considering Knudsen's fundamental contribution to physical acoustics and its influence in present day modelling techniques.

Knudsen published over 100 articles in various scientific and technical journals and I propose to dwell a little on his series of papers on sound absorption in the atmosphere. Using reverberation techniques and the greatly improved facilities at UCLA he made absorption measurements of greater precision than formerly possible but he noticed that the repetition of readings from day to day was poor. In calibrating the reverberation room it was found that the reverberation time above about 2000 Hz depended greatly upon humidity. When moist air from the ocean controlled the relative humidity (greater than 50 per cent) the reverberation time for a 4000 Hz tone was 4 to 5 s while at low humidity, arising from a Santa Ana wind, for the same tone it would be of the order of 2·0 to 2·5 s. This effect was so large that it demanded further investigation since it was at variance with the classical absorption theories of Kirchhoff, Stokes and Rayleigh. These theories attributed the absorption mainly to viscosity and heat conduction and to a minor extent to radiation, and only a negligible effect of water vapour was predicted. Knudsen in common with many other acoustical physicists thought that the water vapour in some way affected the humidity at the boundary surfaces of the enclosure. His first approach was to vary the nature of the containing surfaces of his experimental chamber by painting and then enamelling them, but this was found not to have any appreciable effect on the variation of the reverberation time with humidity.

Knudsen then decided upon making a critical experiment in which he sought to find as a function of frequency at different temperatures and humidities (a) the absorption of sound in air and (b) the absorption coefficients of the boundary surfaces of the room. By using the large reverberation chamber and an adjacent smaller test room he had a system suitable for the purpose since the mean free path would be more than twice as large in the bigger room.

The initial results were published in a paper in *JASA* in 1931 and briefly the conclusions were that the absorption coefficient in air was of the order of 10 to 25 times greater than that calculated from classical theory and that it increased markedly as the humidity of the air decreased. He also found that within the range of his experiments the absorption coefficient of the boundaries remained constant as a function of frequency, temperature and relative humidity. In a paper two years later Knudsen reported an extension of his earlier measurements to dry air and to a range of temperature between $-15°C$ and $55°C$. Besides confirming his earlier measurements these experiments revealed regions of selective absorption at low humidities which have both practical and theoretical interest. It had been suggested by Jeans, Herzfeld and others including Kneser that part of the sound absorption in the atmosphere might occur from the transfer of energy between the translational and either rotational or vibrational forms of energy during molecular collisions.

Professor Kneser cooperated with Knudsen in these later experiments and he proposed a highly satisfactory explanation of many of the experimental results by assuming that a large part of the absorption is attributable to collisions between H_2O and O_2 molecules. The experiment, carried out subsequently by Knudsen in a 2-ft chamber filled with different proportions of water vapour to oxygen, showed quite conclusively that the excessive absorption in air is mostly attributable to an interaction between O_2 and H_2O molecules. He pointed out that if we lived in an atmosphere of oxygen at a temperature of 20°C and with a relative humidity of 26 per cent the high notes of the violin and piccolo would be completely inaudible at 50 m from source. Similar experiments made by Knudsen with nitrogen and water vapour in the frequency range 3 kHz at both 20°C and 55°C indicated that the rate of decay of sound is almost independent of the humidity or temperature at any one frequency. Knudsen was always mindful of the implications of his experiments and in his 1933 paper concluded from his results that for a room with perfectly reflecting walls at a frequency of 10 kHz, temperature 70°F and humidity 18 per cent, the longest possible reverberation time is 0·625 s. Even at a more reasonable humidity of 50 per cent it would be only approximately doubled but this would be reduced in practice if absorption at the walls is considered. He then asserts that the boundaries of a room should be as non-absorptive as possible at the high frequencies, and to obtain the highest acoustical quality possible in all auditoria there should be proper control of humidity as well as of temperature.

Knudsen continued this work and published a paper in 1936 entitled 'The Absorption of High Frequency Sound in Oxygen containing small amounts of Water Vapour or Ammonia' and he extended the frequency range of his measurements to 34 kHz, employing an intensity method. The measurements gave further support to Kneser's theory. Meanwhile in 1935 a general paper on 'The Absorption of Sound in Gases' was printed in *JASA*, being the substance of a presentation at the annual general meeting of the American Association for the Advancement of Science, which gained Knudsen the annual 1000 dollar AAAS prize as the noteworthy paper of the meeting.

It is interesting to note that the follow-up of a problem from an applied field, resulting from observations using improved instrumentation, led to a world-wide experimental and theoretical research programme on molecular phenomena in fluids.

I should have mentioned that a paper in 1932 by Knudsen sought to clear up the confusion that often arose between reverberation and resonance in rooms. He enumerated 8 specific deductions of which two examples will be quoted. He stated that appropriate corrections for the effect of room resonance should be applied to all acoustical instruments which are calibrated in enclosed spaces. Furthermore he stated that the effects of resonance in small rooms used for sound-transmission and sound-absorption tests can be partially overcome by suitable rotating vanes or paddles in the rooms. For the transmission measurements they should be used in both source and test rooms.

We now come to the work which Knudsen and Delsasso were primarily occupied with in their last years, that of acoustic modelling. Knudsen made a number of contributions during the last six or seven years at American Acoustical Society meetings but most of these do not appear to have been published in science journals. His paper in *JASA* of Feb. 1969 was entitled 'Model Testing of Auditoriums'. After reviewing past work on model testing to aid the design of auditoria, he discussed the 2-dimensional model tests developed at UCLA by Leonard in 1954, which are useful for determining approximate optimal shapes in plan and sections. He determined how various parts can be examined separately and progressively assembled in a 3-dimensional model. Examples are given of procedures in designing several multi-purpose auditoria and also of a theatre in the round (*i.e.* 220°) with a thrust stage.

A conclusion made by Knudsen, which confirms the observations of others, is that for certain concert halls which have little diffusion and where most of the absorption is in the floor, the reverberation time is much longer than it is at seat-level. His model experiments also demonstrated that for many rectangular rooms in which there is little diffusion, the absorptive treatment of the floor reduces reverberation and noise more than would the same absorptive treatment of the ceiling. However, with the diffusion of twenty helium-filled balloons (42-in diameter) randomly distributed in the room (with 2-in mineral wool covered floor) the reverberation characteristics were practically independent of the location of the source and microphone.

I am sure from what I have said, even though inadequately representing his life's contribution to acoustics, you will agree that Professor Vern Knudsen was one of the world's most outstanding figures in acoustics, not merely in the limited field of architectural design, but in the fundamental aspects of the subject and its social impact on man.

As a person he was a man of innate goodness and modesty who will always be remembered with sincere affection by those who knew him. It was the Frenchman Jules Renard who said 'the reward of great men is that, long after they have died, one is not quite sure that they are dead', and so it will be with Vern Oliver Knudsen. Although the sound of the hunting horn dies away in the wind, the utterances of acoustic wisdom by Vern Knudsen will be sustained through the passage of time.

12

The Acoustic Design of Multi-purpose Halls

HEINRICH KUTTRUFF

*Institut für Technische Akustik
der Rheinisch-Westfälischen Technischen
Hochschule, Aachen, Federal Republic of Germany.*

When a medium sized community or city is planning a hall for public events, the result is most frequently a so called multi-purpose hall which is not to be used for one specified sort of event or performance but rather for the complete spectrum of purposes for which many people usually come together. The hall is therefore intended to serve for political meetings, for theatre presentations, for sporting events, for concerts of all kinds, for fashion shows and for many other things. There are good reasons for this intended versatility which are mainly of an economic nature and which do not need any more detailed explanation. The architects usually can cope with the situation and they are prepared to find more or less original solutions to the problem to which they are faced so they provide for movable chairs, for a flat floor, for a nice restaurant and so on. The acoustic consultant is in a less happy situation, he is expected to give the hall good acoustics, whatever this means. His question, 'Good acoustics for what?' usually remains unanswered and very often he has to find out by himself what the owner really wants to do with the hall and it is left to him to interpret what is really meant by the words 'good acoustics'. Clearly, the problem is to find acoustical conditions which are favourable for all events to take place in the hall.

As far as reverberation time is concerned there are essentially two opposite positions, the long reverberation time required for orchestral music, in order to make it sound full and smooth and the short sound decay which is necessary for lectures and addresses, since long reverberation affects the intelligibility of speech. But good acoustics is not only a matter of reverberation time. We know that strong early sound reflection increases the useful loudness of sound, which is very valuable for speech, but may mask the reverberation.

Further matters under discussion among acousticians include diffusion, for instance, and quite recently Manfred Schroeder and his co-workers at Göttingen have found out that inter-aural correlation plays an important role in the acoustics—in the subjective acoustics of concert halls and probably for opera houses as well. The latter quantity obviously depends to a considerable amount on the number and strength of lateral sound reflections. That means in a hall intended for musical performances we need strong reflections from the side walls in order to provide for the subjective sensation of space. For the intelligibility of speech, this type of reflection is not so important and in the latter case ceiling reflections may be more useful.

Coming back to the problem of adapting the acoustics of a hall for various purposes, there are virtually three ways to arrive at a solution. The first is to find a compromise with respect to reverberation time and other characteristic parameters. This compromise should take into account the relative frequencies of the various kinds of events. This way is probably the least expensive and has the advantage that nothing has to be controlled or changed after the hall is completed. On the other hand it results in non-optimal conditions for some types of performances and the acoustician should be aware that there will always be some critics of the quality of the work he has done.

The second way is to instal changeable elements such as wall or ceiling sections which can be rotated in order to expose reflecting phases in one position and absorbing ones in the other. This quite expensive method is sometimes applied to broadcasting studios but less frequently to other halls. It can be quite effective in controlling the reverberation time but there is a danger of wrong operation. The modification of this method consists of movable walls by which certain parts of the room volume can be separated acoustically from the main space. This results in a change of reverberation, depending on the volume and the absorbing area in the separated portions. Additionally the temporal and the directional distribution of sound reflections are altered by the separating walls.

The third way is to give the hall optimum acoustical conditions for one kind of presentation and to try to adapt it to the others by electro-acoustical means. If we make the natural reverberation time relatively long, favouring orchestra and opera concerts, we can try to overcome the reverberating sound field during speech presentations by suitably designed loudspeaker installations. This can be done by carefully adjusting the directional characteristics of the speakers so as to direct the sound strictly to the audience and to avoid, as far as possible, the excitation of reverberation. I shall put it in a different way, every listener should be placed inside the reverberation distance with respect to one or more loudspeakers. In my own opinion the possibilities of this method are not yet exhausted. The opposite approach starts from a relatively short decay time as this is favourable for all uses of the hall in which speech is involved. For musical performance, the reverberation time is increased by electro-acoustics. The result of this procedure is most satisfactory of course if the required increase is not too high. Increasing the decay time can be effected by making positive use of acoustical feedback, of course with all necessary precautions. The most prominent system of this kind is known as the assisted resonance system.

My own experience is restricted to another type of system in which the additional reverberation is produced by a separate reverberation chamber or an equivalent device and is subsequently transplanted to the original room by many loudspeakers. In this case it is important to simulate, not only the duration of sound decay but also its directional distribution and of course the temporal succession of direct and reverberated signal as it might be in naturally reverberant sound fields. In the following I shall describe one example for each of these possibilities by referring to three multi-purpose halls built during the past decade and I shall report on the experience gained with them as far as possible. All these halls are located in German cities of medium size.

The first of these is the so-called 'Forum of Leverkusen' which is a city of about one hundred and twenty thousand inhabitants, not far from

Fig. 1. The Forum at Leverkusen.

Cologne and it is the site of a large chemical company, which is the most prominent employer and tax-payer of the region. Figure 1 shows the buildings situated fairly close to a highway and consisting of a large hall, several smaller rooms for chamber music, lectures and so on, and of a large foyer used also for exhibitions, popular events and carnival parties and so on. The large hall, a multi-purpose hall of course, has a volume of 6500 cubic metres corresponding to 230 000 ft^3 and can seat nearly 1000 persons. In the sections in Figs. 2 and 3 we see a large stage house which is fitted out with all technical installations required for theatre presentation including an orchestra pit for about 60 musicians, 2 galleries for stage-lights and a

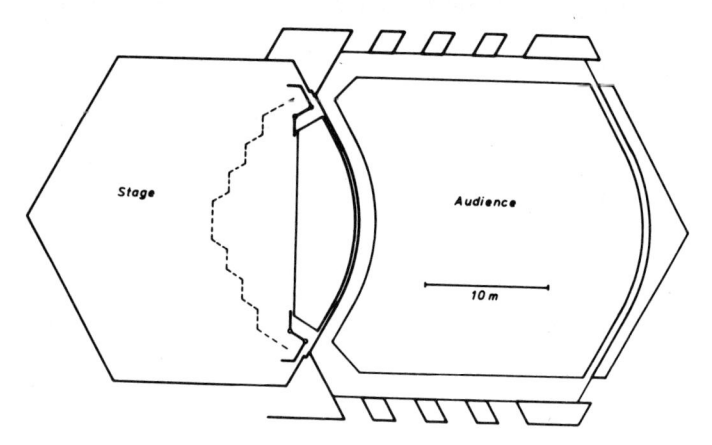

Fig. 2. The Forum at Leverkusen—plan.

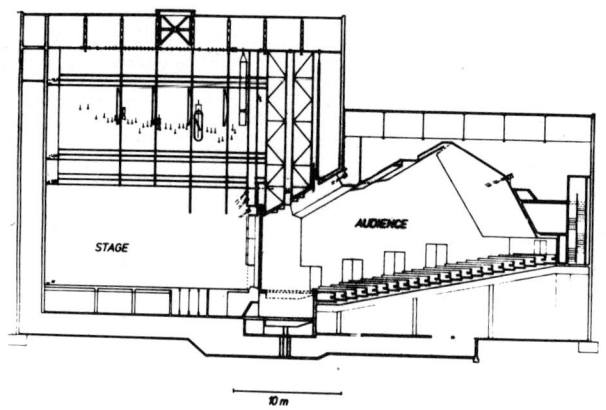

Fig. 3. The Forum at Leverkusen—cross sectional elevation.

control room at the rear. Of course there are no reflecting areas in the stage region, neither on the side walls nor on the ceiling, thus it was desirable to use each square metre of the remaining ceiling to produce useful sound reflections towards the audience. For orchestra concerts there is a movable stage enclosure. The side walls are essentially parallel and are lined with wooden panels backed by air.

Since the shape of the hall did not appear to be free of acoustical problems, thorough acoustical model experiments had been carried out during the planning phase which revealed the danger of audible echoes due to the oblique parts of the ceiling. This section got a sound absorbing treatment by perforating the faces and making them of Rockwool. The same was done with the concave curved rear wall. The side wall lining consists of small sections, which are directed irregularly into different directions in order to avoid flutter echoes and to produce somewhat diffuse reflections. Apart from these, the walls act as low frequency sound absorbers. The floor is covered with a relatively thick carpet. The resulting reverberation time of the empty room with the steel curtain closed is plotted as the upper graph in Fig. 4. The lower graph shows reverberation time as calculated from the upper graph, taking into account the absorption increase caused by the audience as measured in a large reverberation chamber. At medium frequencies the decay time is between 1·3 and 1·4 s which is quite favourable for theatre or chamber music. It is still not bad for lectures or for opera but not so good for concerts. This lack of reverberation is tolerable however

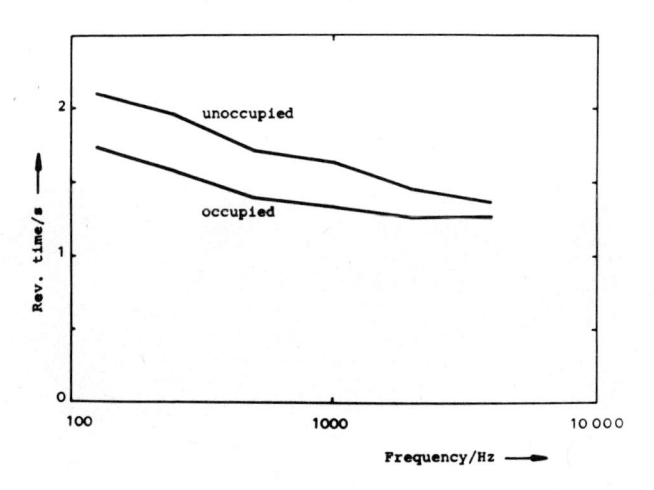

Fig. 4. The Forum at Leverkusen—reverberation time.

since only 12% of all presentations of the past season have been orchestral concerts, with 42% being theatre plays without any music and 23% being opera performances. This indicates that the point of compromise had been chosen so as to create favourable conditions for the majority of the events as they actually take place. The public has accepted the hall favourably although some complaints were uttered shortly after the opening in 1970 but these complaints did not concern a maladjustment of reverberation but rather a lack of loudness at certain places.

The next hall I am going to describe is the Stadthalle of Braunschweig. In former times, Braunschweig was the residence of a Duke, for this reason it has a long standing cultural tradition. Nowadays it has about 250 000 inhabitants. In the sixties it was decided to erect a new and modern cultural centre comprising various rooms, among them a large hall for various purposes but with some emphasis on orchestra concerts for which no suitable place had been available so far. The floor plan of this hall is presented in Fig. 5. It has the shape of a regular triangle with truncated corners. In one of these corners there is a stage, the area of which is 450 m^2 at maximum, parts of it can be lowered so as to form an orchestra pit. In the remaining corners there are elevated annexes or terraces with additional space

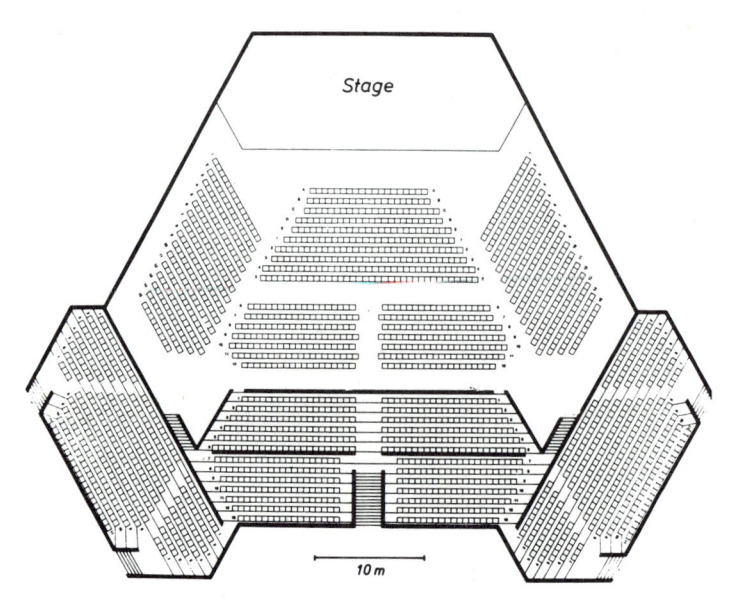

FIG. 5. Stadthalle at Braunschweig—plan.

for audience. These balconies or terraces have volumes of about 900 m³ and 325 seats each. The total volume of the hall amounts to 18 000 m³ corresponding to 630 000 ft³. The total seating capacity is nearly 2200. The balconies can be separated from the main hall by folding walls. The ceiling of the hall is lightly curved in its general shape, in detail however it is covered by a great number of elements (see Fig. 7). These elements have some resemblance to clipped pelmets and are each about 7 m long and 1·5 m high. They are supposed to act as acoustical scatterers in order to destroy sound focusing by the ceiling. Furthermore each of them is a techni-

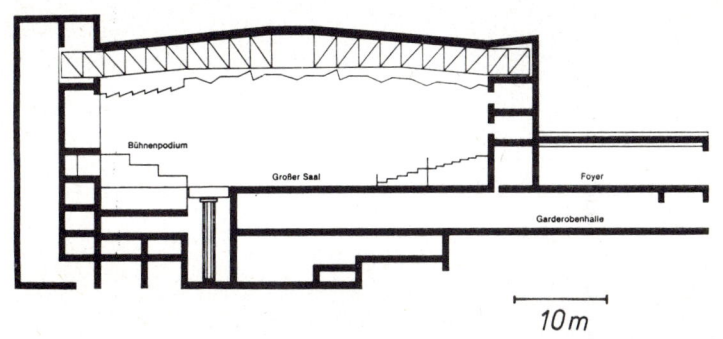

FIG. 6. Stadthalle at Braunschweig—cross sectional elevation.
Key: Grosser Saal = Large hall; Foyer = Foyer; Garderobenhalle = Cloakroom; Bühnenpodium = Stage.

FIG. 7. Stadthalle at Braunschweig—interior view.

cal unit containing lights for the stage and the audience, loudspeakers and air conditioning installations. The side walls of the room are lined with wooden panels which are 2 cm thick and are backed by an air space of 12 cm thickness. This space is partially filled with Rockwool. In the stage area the ceiling consists of parallel strips lined in such a way that the sound is directed towards the audience.

The reverberation time of the occupied hall both with open and with closed rear balconies or annexes is shown in Fig. 8. With closed balconies the decay time is about 1·6 s at medium frequencies increasing up to more

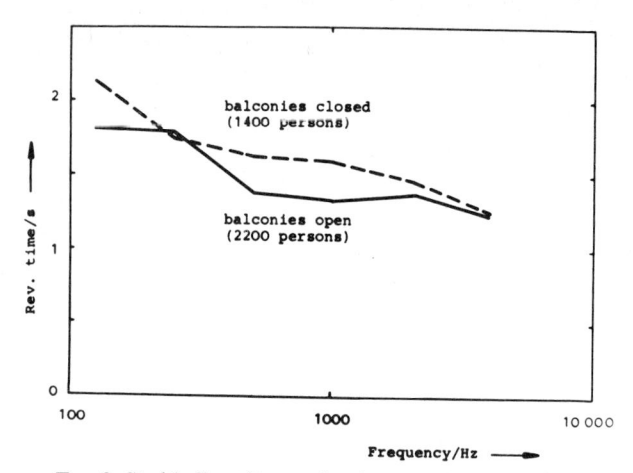

FIG. 8. Stadthalle at Braunschweig—reverberation time.

than 2 s at about 100 Hz. If the balconies are open and all the seats occupied, the decay time drops to 1·3 to 1·4 s at medium frequencies and is only 1·8 s at low frequencies. This reduction of reverberation time is due to a much smaller volume per seat in the balconies than that of the main room. To a first approximation the balcony openings can be considered as perfectly absorbing areas, at least at higher frequencies. For orchestral concerts the halls should apparently be used with closed balcony openings, not only for the value of a longer reverberation time but also for increasing the number of strong lateral sound reflections. The 1450 seats to be occupied in this situation are sufficient for a city like Braunschweig. For more popular events, or for meetings if the full capacity of the hall is employed, a lower decay time is adequate. Incidentally, the public address system is not optimal in my opinion since it employs distributed loud-speakers which are fed by identical electrical signals. The system ensures sufficient loudness and good intelligibility of speech but the listener localises the sound source under these conditions. But apart from this the hall is appreciated as a fine concert hall as well as a hall for all types of meetings.

The most interesting of my examples is probably the Jahrhunderthalle Hoechst seen in Fig. 9—a building which was erected by the Farbwerke Hoechst of Frankfurt on the occasion of its centenary in 1963. This hall has been described already elsewhere but it seems worthwhile to do this once

FIG. 9. Jahrhunderthalle Hoechst.

more and to report on the experience gained with it and with its electroacoustical installations. This hall looks quite impressive outside as well as inside. It is essentially a huge spherical shell made of concrete and covered with blades of plastic material. The height is 25 m and it has an extrapolated diameter of 87 m. Actually there is a side wall consisting of a glass cylinder with a height of 6 m and a diameter of 78 m. The total volume is 75 000 m^3, corresponding to nearly 3 000 000 ft^3. Unfortunately, impressive architecture sometimes causes considerable headache to the acoustic consultant, especially if the building is not intended to be just a monument but to be used for practical things. In the present case, this object is a nearly-all-purpose hall used for boxing matches, for string quartet presentations, for share-holder meetings, for orchestral concerts of course and for all kinds of popular events. It is the largest hall in the area around Frankfurt, including Frankfurt itself.

Mainly for reasons of geometrical acoustics, it was decided at the very beginning of planning to make the shell and the side wall cylinder as dead as possible from the acoustical stand point. Natural sound reflections are produced only by the rigid stage enclosure and by a pattern of suspended panels above the stage and in ascending rows above the audience area. This is clearly not sufficient to create acoustical conditions favourable for musical presentations. For this reason, an electro-acoustical installation was installed with a three fold task, namely to amplify the direct sound which is clearly necessary for lectures, addresses and so on but also for small musical ensembles, for string quartets for instance. The second purpose of the system is to simulate strong and distinct reflections which are normally produced by side walls not present here in this case and thirdly the system has to produce reverberation which is necessary for all musical performances.

Starting with the room acoustical treatment (see Fig. 10) the dome is covered on its inner side with a layer of Rockwool 15 cm thick on average which consists of two layers with different densities. Behind the Rockwool there is an air space of 20 cm thickness. On the room side the Rockwool is protected by a grid of thin strips. The glass cylinder (which can be seen in Fig. 11) that is the side wall is made absorbent by a deeply folded curtain made of relatively dense and heavy material. The average distance

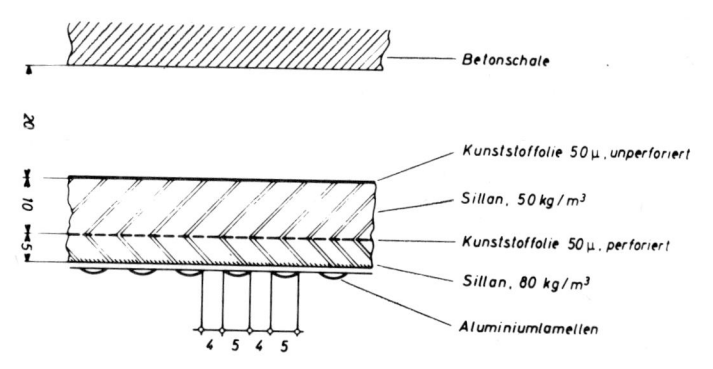

FIG. 10. Jahrhunderthalle Hoechst—acoustic treatment.
Key: Betonschale = Concrete shell; Kunststoffolie 50μ unperforiert = Unperforated plastic skin 50μ; Kunststoffolie 50μ perforiert = Perforated plastic skin 50μ; Sillan, 50kg/m^3 = Sillan, 50kg/m^3; Aluminiumlamellen = Aluminium sheet.

between this curtain and the glass wall is 70 cm, therefore the absorption is high even at low frequencies. The stage enclosure has undergone several modifications since the hall opening in 1963. Its fixed areas are the interior parallel faces of the two giant concrete pillars. Various experiments have been carried out with stage enclosures, smooth ones and corrugated ones

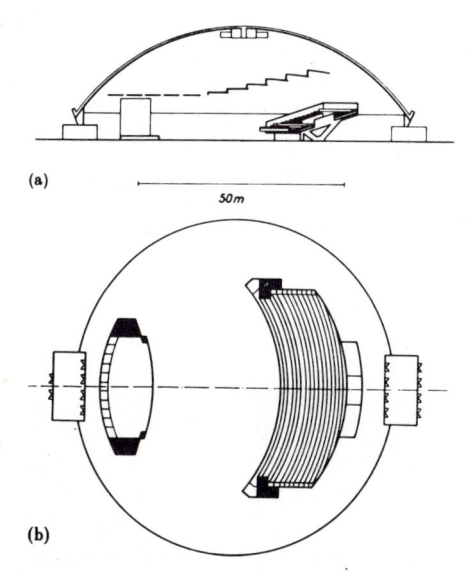

FIG. 11. Jahrhunderthalle Hoechst—plan and sectional elevation.

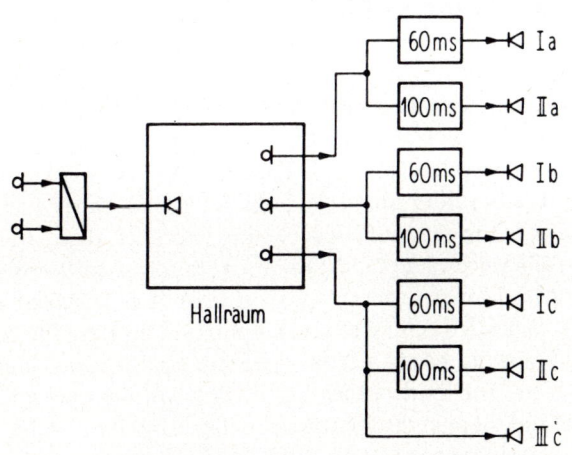

FIG. 12. A simplified block diagram of the reverberation system.
Key: Hallraum = Reverberation chamber.

and right now consideration is being given to constructing a new enclosure which incorporates the experience gained with its predecessors. Figure 12 shows the hall as viewed from the stage. In reality the suspended ceiling elements are not curved of course, but flat. At full occupation the hall seats up to 3500 persons, depending on the number of chairs which are set up. For concerts, theatre plays and so on, however, the size of the audience is restricted to about 2000.

During the first reverberation measurements, it was a great surprise to see that despite the heavily absorbent treatment of almost all surfaces, the natural reverberation time with about 2000 persons present was about 1 s and that decay curves looked as regular and normal as seen in text books on acoustics. They did not show any curvature or double slope. First their natural reverberation time was considered good for speech. The system employed for increasing the reverberation time for other purposes is shown in a simplified version in Fig. 12. The sound signal is picked up by several microphones close to the sound sources—that is on the left hand and after some mixing it is replayed in a reverberation chamber by a loudspeaker. This chamber has a volume of 144 m^3 and is located in the basement of the building. Its reverberation time is adjusted by specially designed sound absorbers to a bit more than 2 s and is virtually frequency independent. The reverberated signal is picked up once more by three microphones delayed by suitable delay times and fed to three different groups of loudspeakers. The three output voltages of the microphones can be considered mutually incoherent which is very important with respect to the subjective impression caused by the radiated sound signals.

The most crucial point of this system is the arrangement of the loudspeakers. When we started with this work, there was no experience at all in this matter and therefore it was decided to distribute 60 loudspeakers in a semi-circle, in the upper part as well as in the lower part of the dome surrounding the audience (see Fig. 13). The idea was to produce something like a diffuse reverberating sound field. An additional thirty loudspeakers have been mounted in the suspended ceiling elements, these are the open boxes in Fig. 13. Since these loudspeakers were too small to produce low frequencies, special low frequency speakers have been installed above the centre of this suspended ceiling at several points. These are the boxes with a dot inside. The solid boxes are loudspeakers used for the simulation of distinct reflection. The loudspeakers in the first row are operated at delay times of 30 ms and those in the 4th row are delayed by 70 ms. All delays were produced by rotating magnetic discs.

Coming back to the reproduction of the reverberated sound signals in

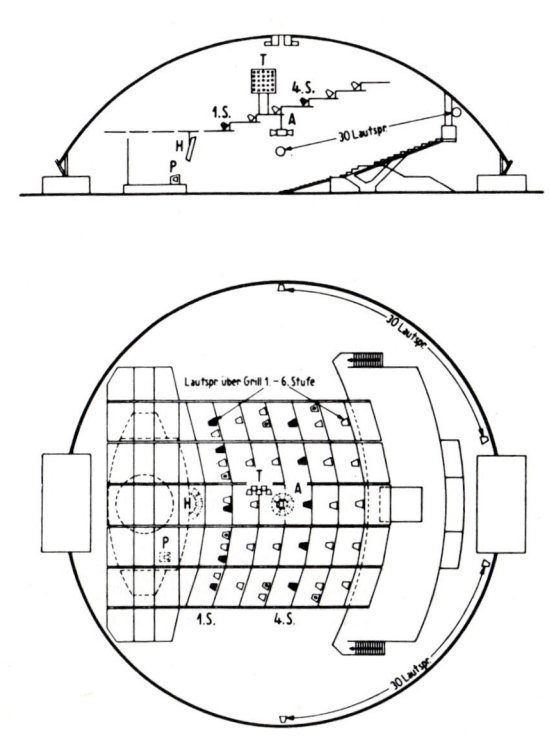

FIG. 13. Loudspeaker arrangement in Jahrhunderthalle Hoechst.
Key: 30 lautspr. = 30 loudspeakers; lautspr. über grille = loudspeaker above grille.

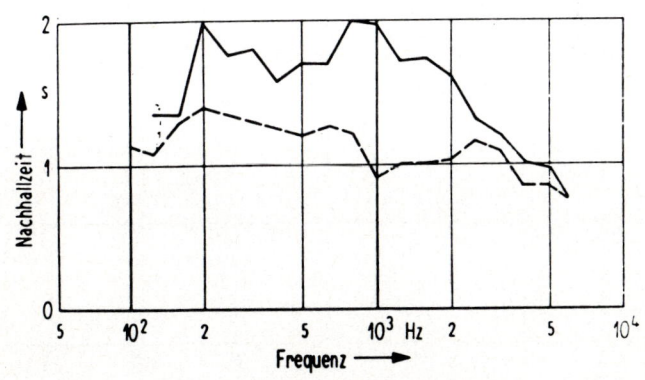

FIG. 14. Reverberation time in Jahrhunderthalle Hoechst with (———) and without
(– – – –) electro-acoustical system.
Key: Frequenz = Frequency; Nachallzeit = Reverberation time.

the hall, it is quite clear that the amplification in the reverberation channel is very critical. If it is too low no increase in the decay time can be noticed, if it is too long, on the other hand, the sound quality becomes degraded on account of correlations due to acoustical feedback. In Fig. 14 the reverberation time of the hall both with and without the reverberation system is plotted as a function of frequency. These measurements were taken during a test concert with 2300 persons present. According to this figure the reverberation time can be raised by the system from about 1·1 to 1·8 on average.

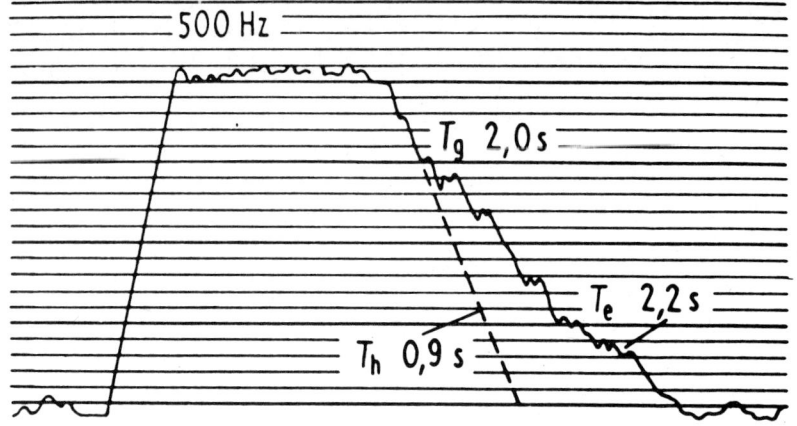

FIG. 15. Reverberation course with electro-acoustical reverberation system in operation.

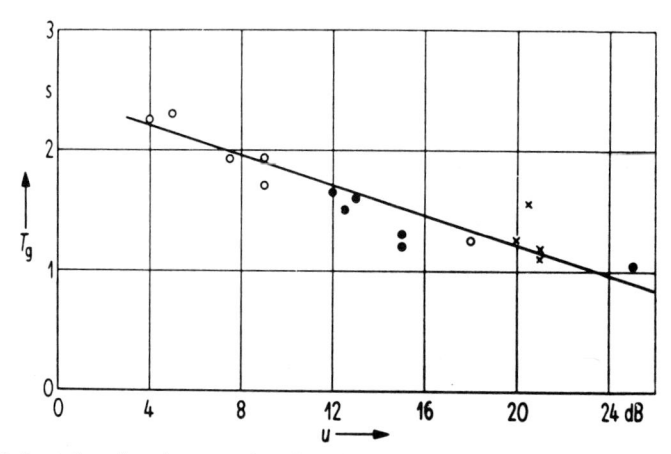

FIG. 16. Reverberation time as a function of the ratio of sound energy to reverberated sound energy.

The drop at low frequencies is due to the poor low frequency performance of the loudspeakers utilised at that time. Of course the reverberation curves are no longer straight lines when the system is in operation as can be seen from Fig. 15 here. It has turned out that by a proper mixing of the reverberated signal with an unreverberated electrical signal the perceived decay time, that is T_g could be made variable within certain limits. This is illustrated by Fig. 16. In this diagram the abscissa is the level difference between the total sound pressure at a certain place and the sound pressure level of the direct sound, both at steady state conditions. The reverberation time plotted on the ordinate is that measured at the corresponding setting of the amplifier. It is evident that reverberation time can be varied continuously, or what is heard as reverberation time can be varied continuously by varying this level difference, that is by varying the relative contributions of the reverberated and the unreverberated sound field components.

During the past decade several changes and improvements have been made with this electro-acoustical system. The sound signal which is to be amplified without any reverberation is radiated by four rows of loudspeakers now suspended in the ceiling delayed by 30, 50, 80 and 100 ms. Each loudspeaker consists of a box with 3 units. For reverberating sound signals picked up on the stage a reverberation plate was used in addition to the reverberation chamber. This has the advantage of increased eigenfrequency densities at low frequencies as compared to that of the relatively small room. This fact is a consequence of the dispersion of bending waves as they are propagated within reverberation plates. Quite recently both the chamber and the reverberation plate have been replaced by an electromechanical reverberator employing transverse waves on metal springs. Reverberators of this type have long been known in principle but they have been greatly improved recently by inserting irregularities into the springs and by simulating the properties of natural reverberation by complicated and ingenious feedback loops. So far these reverberators yield, reportedly, the best results concerning the quality of reverberated sound. For the reproduction of the reverberated signals in the hall only loudspeakers in the suspended ceiling are utilised. The original ones have also been replaced by hi-fi three-way boxes each handling an electrical power of 100 W.

This reverberation system is applied to all performances of opera, chamber music, orchestral music, for chorus, concerts and ballets, that is for all presentations with music in any form and all this has been done now for 10 years. As an example I would like to mention the presentations of the 1972 season. In this year there were 3 chorus concerts, 9 orchestral concerts, 2 chamber music presentations, 2 opera and 2 operetta performances,

5 ballet presentations, 4 test concerts and 10 theatre plays. For all these events, except the theatre plays, the artificial reverberation system was in operation. This indicates that increasing the decay time in this way is a really practical method that is no longer in the phase of mere experimentation but has become a legitimate and useful tool in room acoustics. I am sure that in the long run the use of electro-acoustical aids will prove to be the best and most versatile way to settle the acoustical problems of large multi-purpose halls. Nevertheless, I am aware of the fact that much experience still has to be collected in order to make these techniques as reliable and controllable as for instance the illumination, or the air conditioning of a large hall.

13

Different Distributions of the Audience

LOTHAR CREMER*

Institut für technische Akustik,
Technische Universität,
Berlin, Federal Republic of Germany.

If I might assume that many members of the audience here are architects and engineers and not acoustics consultants, I shall refer to a problem which is primarily of visual concern and thus important architecturally. However, in doing so I will explain the acoustic considerations which are connected with it. The question which must be answered at the very beginning of each project involving the design of an auditorium where a performance is to be seen and heard is that of how a given capacity of audience is to be distributed.

In referring to the history of theatres, the earliest well-developed solution was the amphitheatre. Figure 1 shows not the earliest, but the best maintained example, the theatre of Epidaurus in the North-West of the Peloponnese, before its restoration.

FIG. 1. The theatre of Epidaurus—before restoration.

* Present address: Immanuel Kant Strasse 12, 816 Miesbach, Federal Republic of Germany.

It may be that such a beautiful and historical background influences also the judgement of the acoustics. But there is no doubt that you can hear the speech of the actors with sufficient volume and a remarkable distinctness without electro-acoustical amplification. Certainly this requires as a prerequisite two features which have now come to be more and more rare, a quiet landscape and a quiet audience, and we may doubt that both will continue in the future. As soon as we make use of the electro-acoustic facilities of our time the practical interest in the problem of the special acoustical qualities of the ancient theatres disappears. Sometimes we assume that it was as a consequence of the development of electro-acoustics that acoustic signals could be addressed to a large audience but the theatre of Epidaurus was built initially for 6000 people (Fig. 2, inner circle) about 300 B.C. and enlarged for 14 000 people about 100 years later. The ancient Greek architects had to solve the problem of large auditoria but without electro-acoustic means and they solved it. This has given rise to the assumption that they must have had especial knowledge of acoustics but we have no information about such a knowledge. Vitruvius's *Ten books on archi-*

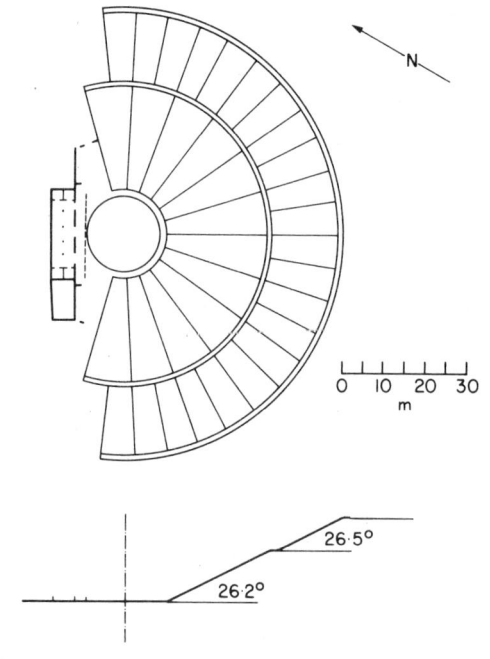

FIG. 2. Plan and sectional elevation of theatre of Epidaurus.

tecture, often quoted as an authority, were written about 300 years after the building of the inner circle of Epidaurus; we can scarcely assume that an author of the twenty-third century would give an adequate explanation of our present way of building without our present literature.

In any case we may assume that they intended to have good visual conditions for as many spectators as possible. This was one of the reasons for the high steps from row to row, which furthermore corresponded to the normal seat-height. The rather steep slope of the audience (seen in the section of Fig. 2) offers the acoustic advantage that the sound from an actor standing before the stage-house meets even the highest part of the audience at a rather large angle ε between sound-ray and audience-line. How important this angle is, becomes evident when an actor speaks at the front of the so called orchestra, the circular place before the stage-house. In this situation, the sound approaches the audience at grazing incidence with the result that loudness and clarity become remarkably worse.

In another Greek theatre, the so called 'Herodes Atticus' in Athens (Fig. 3), I heard a performance of Tchaikovsky's ballet *Swan Lake*. The sound of the first violins, placed near the barrier as usual, appeared to be very 'thin' in the middle of the audience but became pleasant near the high wall behind the stage.

It is not only the angle of incidence for the direct sound that profits from the steep slope in Epidaurus, this is even the case for the sound reflected

FIG. 3. The 'Herodes Atticus' theatre in Athens.

from the orchestra which supports the direct sound on account of the short time-delay. The late Professor Canac, an enthusiastic investigator of the ancient amphitheatres, even assumed that the Greek architects took this into account by the choice of their dimensions (Fig. 4).

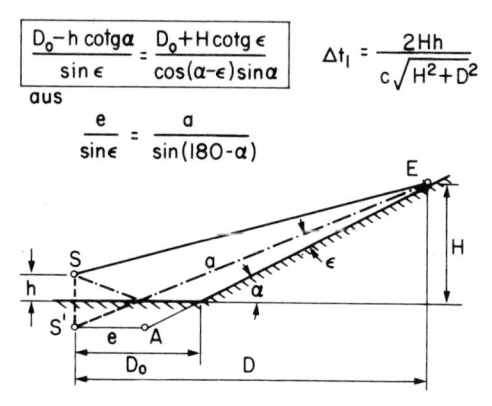

$$\frac{D_0 - h \cot g \alpha}{\sin \epsilon} = \frac{D_0 + H \cot g \epsilon}{\cos(\alpha - \epsilon) \sin \alpha} \qquad \Delta t_l = \frac{2Hh}{c\sqrt{H^2 + D^2}}$$

aus

$$\frac{e}{\sin \epsilon} = \frac{a}{\sin(180 - \alpha)}$$

FIG. 4. Geometrical relationships in theatres according to Canac's canonical formula showing time-delay between direct sound-ray and ray reflected from the floor (c = velocity of sound).

Without going into details—because the exact solution is rather complicated even if we regard the audience as a homogeneous layer of absorbing material—I might mention two reasons for the importance of the above mentioned angle ε. Firstly the head-plane of the audience is a plane where the sound pressure is reflected with nearly opposite phases at grazing incidence. So the heads are nearly in a pressure node and more so as ε approaches zero and thus one might say, as a first approximation, that the resultant pressure increases with ε. The second reason, which for the layman is easier to understand, consists of the fact that the audience absorbs sound, especially on account of the porous fabrics of their clothes, and therefore waves propagating tangential to the audience (so called surface-waves) lose energy permanently. Both have been studied theoretically and experimentally by several authors, among them by the late Professor von Békésy. The sound pressure drop over the audience at grazing incidence therefore is often called the Békésy effect by acousticians.

Besides this advantage for the direct sound and for the sound reflected from the orchestra and also from the stage house, both of which follow with a short time delay and support the direct sound, (Vitruvius speaks of 'consonant' sound), the ancient amphitheatres are, when occupied by the audience, characterised by the lack of delayed reflected sound-rays (Vitruvius speaks of 'dissonant' sound) although the circular shape of the rows and higher steps or of the rear wall are acoustically dangerous on account of the focusing of the sound-rays reflected from these surfaces, but these rays are mostly directed against the open air above.

This may not only be concluded from the drawings, to-day it can also be proved experimentally by recording so called echograms (Fig. 5), even in the empty theatre, if the loudspeaker sending a tone-pulse is placed near the stage-house and the microphone is placed near the middle of the audience. The delayed small echoes come from places where no reflection happens in the occupied theatre. But is the presence of a few supporting reflections and the lack of delayed ones the only secret of the praised ancient theatres? It was to investigate this question that my colleagues Dr. Kürer

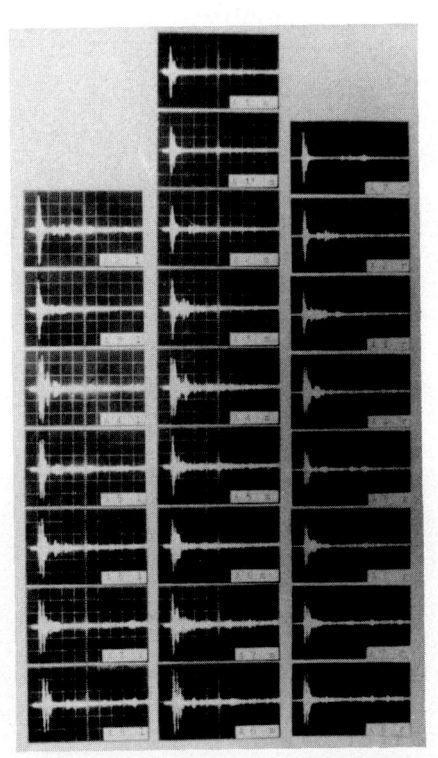

FIG. 5. Echograms in the empty theatre of Epidaurus.

and Dr. Plenge and myself travelled some years ago with the necessary equipment to the Epidaurus theatre. It seemed possible to me that the funnel shaped air space of the theatre tends to produce an air-movement in the direction from the actor to the audience with a vertical increase of flow; but this would give this air-space a lens-character which curves the sound-rays towards the audience. At least this would once more increase angle ε between the direct sound and the tangent to the audience (see Fig. 6). On the other hand, I estimated this effect to be very small and its advantage would have to be proved in the occupied theatre but we had no chance to make adequate measurements during the performance.

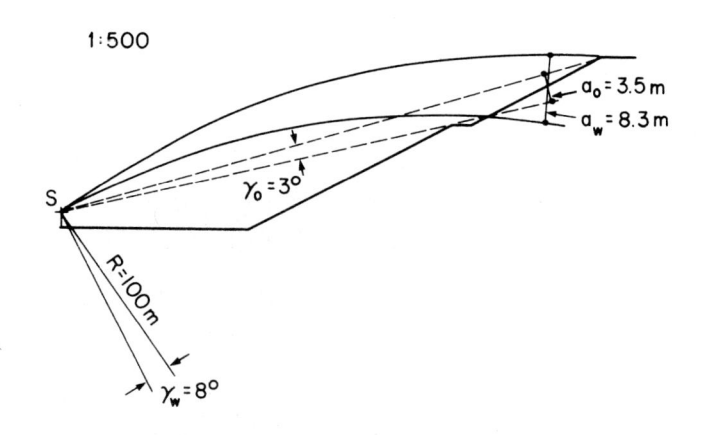

FIG. 6. The effect of air movement upon the distribution of sound at Epidaurus.

However, we could make very careful echograms and measure the first peak in the empty theatre in the early morning and in the late evening whilst the tourists slept. Measurements in the open air are much more difficult than they look. Even if you regard the situation as windless there is always enough movement in the air to give the results a noticeable spread. Figure 7 shows the corresponding cumulative distribution in this case and the greater spread for a rather small wind between 0 and 6 m/s only. If we plot the mean values of this distribution over the distance r to the sound source we find that the sound-pressure, at best, decreases with r^{-1}, as is to be expected under ideal conditions, but in general, measurements in the open air show a large decrease with distance. So this result means that the

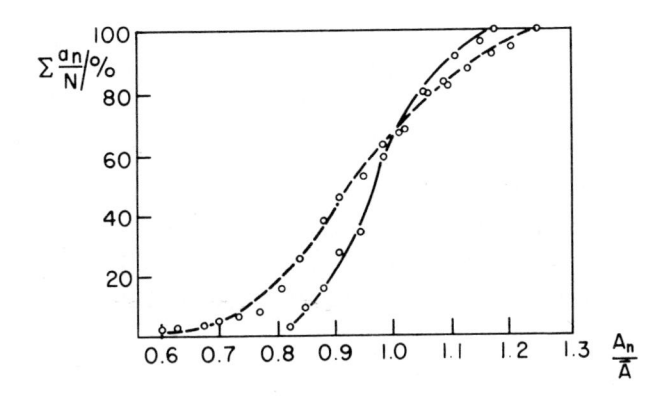

FIG. 7. Epidaurus—cumulative distribution of relative amplitudes of direct sound at a central seat in front of the diazoma.
(———) Gaussian-tonepulse 2kHz 2ms, no wind.
(– – – – –) Gaussian-tonepulse with wind speed 0–6 m/s from different directions.

FIG. 8. Bayreuth Opera House.

sound propagation from the actor to the audience was a bit better in Epidaurus than expected under normal conditions, but no pronounced lens-effect was observed; nevertheless the funnel-shape may have its advantage protecting the air-space against wind-distortion. (But we were confronted with an acoustical mystery I cannot explain. Before the performance started the cicadas in the woods round the theatre made a tremendous noise and then stopped as soon as the performance began.)

From the architectural point of view the most important aspect of the amphitheatre is the social equality of all spectators. This aspect and the good visual conditions were the only reasons that Richard Wagner and his architect returned to this distribution of seats when they built the Bayreuth Opera House (Fig. 8). In the final event this distribution could not be achieved completely since the King, Ludwig II, was the sponsor of Wagner's work and requested that the rear wall be covered with boxes above the amphitheatre for the high society and, above these, for their attendants, but this was acoustic luck. In another theatre, the Prinzregenten Theatre in Munich which was built at the beginning of this century and which was regarded as a copy of the Bayreuth Opera, these higher boxes were not installed. The effect was a double reflection at the ceiling and this part of the rear wall which on account of its cylindrical shape focused an unpleasant echo to the stage. In Bayreuth the higher boxes interrupt this sound-path by absorption. Also, in another way, Bayreuth avoids a difficulty which may occur if one transfers an open air amphitheatre to a closed hall. The fan-shaped ground-plan with its oblique side-walls creates reflections which are useful for people sitting in the immediate neighbourhood only. This again is the case in the Prinzregenten Theatre. In Bayreuth the side-walls are interrupted by projecting cross-walls similar to side-scenes. (Nothing is to be found in Wagner's papers about acoustical reasons for all these rich details of the design.)

The diffuse reflections of the side-walls are a predominant feature of the Baroque theatre with its balconies, boxes and stuccoes. For example, Fig. 9 shows the interior of the Kurfurstliche Oper in Munich which was later named after its architect Cuvillies. The ornamentation of it was removed during the war, so it could be restored in its original form, except for a small break in the highest gallery, needed for the spotlights, and for a picture on the ceiling.

In the Baroque theatre the distribution of seats is totally different to that of the amphitheatre. It was at least as important to see the king's box from everywhere as to see the stage and the situation of a seat with respect to the king's box was a social classification for its possessor. This continued to be the usual opera-house style, in spite of its visual disadvantages for at least one third of the spectators, even in Wagner's time, and when the National Theatre in Munich had to be rebuilt after the war it was decided to do it in its traditional form. This not only meant some places with visual restrictions, it also included an acoustical risk. The National Theatre has the usual curved horse-shoe shaped ground-plan with its focusing areas near the sidewalls. Because of the strict preference for specific stylistic requirements Helmut Müller and I, who were responsible for the acoustics, could not introduce any diffusing elements on the walls to avoid this.

Fig. 9. 'Kurfurstliche Oper' in Munich.

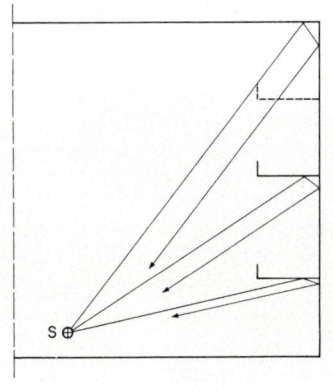

Fig. 10. Sound ray paths between source and balconies.

Our colleagues Vilhelm Jordan and Cyril Harris must have been more successful when they were consultants for the New Metropolitan Opera in the Lincoln Centre in New York. Although there the situation would have been open for a totally new theatre type preference was again given to the Grand Opera Style with its balconies and boxes, but the ground-plan is rectangular and the breast-work shows large diffusing elements. In another respect the construction also merits consideration. The height of the different balconies increases towards the ceiling, so the double reflections from the wall and ceiling of each balcony can bring the sound back to the floor with favourable time delays between them. Otherwise the highest balcony would have interrupted this ray-path (see Fig. 10, dashed line).

Employment of the rectangular shape was the usual modification of seat-distribution adopted for the concert hall of the last century. Figure 11 shows as an example the old Philharmonic Hall of Berlin which was destroyed during the war. Besides the requirement for unrestricted visual lines and for not more than two balconies (for instance in Boston) the concert halls so conceived had, as a result, more volume per seat and therefore longer reverberation times than the opera houses which they copied. In both auditoria the spectators on the balconies have the advantage that the direct sound arrives there without any grazing over the heads of people sitting in front of them.

In the opera houses, for visual reasons, at least a small slope for the audience in the stalls was required but, since many of the concert halls

FIG. 11. The old Philharmonic Hall of Berlin (destroyed during World War II).

were used also for dancing and banquets, they have a horizontal floor and consequently direct sound has to propagate to the last rows with a very small angle between ray and audience-plane. So we have there just what was avoided in the old amphitheatres. Certainly the audience get many reflections from the ceiling and the side-walls but it is still a disadvantage that the direct sound is weakened more than is necessary.

Bearing in mind the two extremes of distribution of seats, the amphitheatre and the balcony-theatre, we may now consider the new Berlin Philharmonic Hall (see Fig. 12), most noted for the peculiarity that the

FIG. 12. The new Berlin Philharmonic Hall.

orchestra is surrounded everywhere by the audience. This was the primary social idea of the architect, the late Hans Scharoun, but this idea is not as new as is generally assumed. In nearly all older concert halls the seats for the choir are sold to the public if they are not needed for the performance. In the Berlin Singakademie and in the Concertgebouw of Amsterdam these areas are very large. But in the Berlin Philharmonic Hall there are also seats for the audience behind the choir and plenty of seats on both sides of the orchestra. This has its advantages; for instance one avoids in a given audience (in Berlin 2300 people) very distant seats, whilst the distant seats gain from having favourable reflections of the ceiling close by. It has, however, its disadvantages; for instance not seeing the singer's face from the seats behind him. Therefore at the start of my consultancy, I tried to dissuade Scharoun from this idea, but he insisted on it and promised to do everything to make his idea viable.

One of the ways of achieving this was by frequent interruption of the

156 *Auditorium Acoustics*

audience area by higher steps. There were two reasons for this, Firstly, it should avoid too much grazing propagation of the direct sound over many rows of people sitting in front of a spectator. Even if we do not get a larger angle ε in this way, the first rows of each terrace are supplied with unaffected direct sound waves and the detrimental effects of interference and absorption mentioned above need several rows to become effective. Sec-

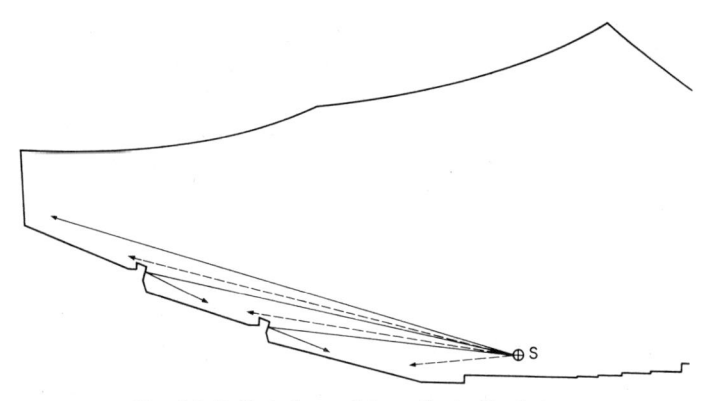

FIG. 13. Reflected sound from the inclined steps.

FIG. 14. Audience splitting in the New Berlin Philharmonic Hall using the 'vineyard-step' principle.

ondly, for the back rows of each terrace the inclined steps supply favourable reflected sound (see Fig. 13). Generally the monotonic absorbing area of the audience is in this way mixed with reflecting areas.

This step-principle had already been found acoustically successful in the design of the Mozart Hall in Liederhalle at Stuttgart by Abel and Gutbrod. In the wine-country around Württemberg the steps were called 'vineyard-steps'. The vineyard-steps not only have an acoustical advantage, they seem to bring the more distant terraces nearer to the orchestra and give the distribution of the audience a new social aspect, especially if this splitting in 'blocks' is solved as ingeniously as Scharoun did in Berlin (see Fig. 14).

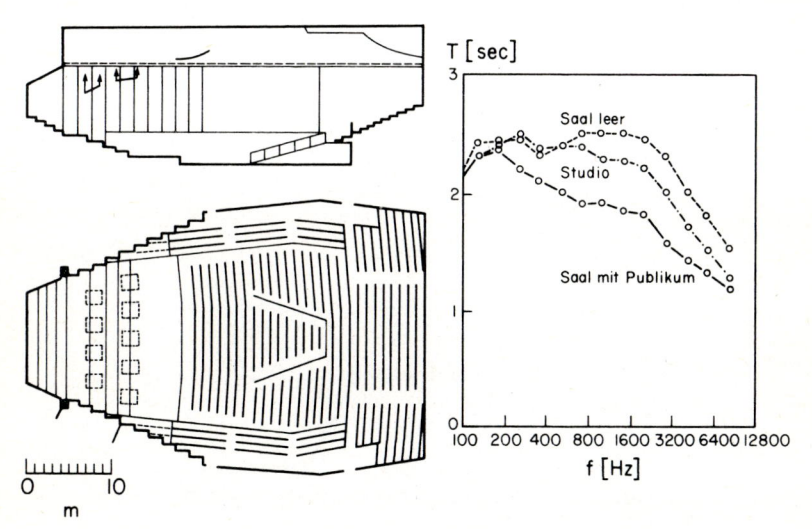

Fig. 15. The large hall of the Berlin Broadcasting SFB.
Key: Saal Leer = Empty hall; Saal mit Publikum = Hall with audience.

The conductor Dorati characterised the situation to me with the words, 'You may look in any direction and always find a group of people whom you may regard as the proper listener, you have made the music for.' The auditor is neither immersed in an innumerable multitude as in the ancient amphitheatres nor classified as in the Royal Operas. I hope professors of sociology will agree that neither is adequate for the society we have in mind, but distribution in blocks is acceptable.

The vineyard-steps in the Berlin Philharmonic also partially offer favourable lateral reflections. The benefit of these reflections was proved especially in a hall of the Berlin Broadcasting SFB (see Fig. 15). Here a wedge-shaped area between ascending slopes over a horizontal floor was provided for television purposes. Although there one sits behind a lot of occupied seats on a horizontal floor, the reflection of the wedge-walls follows the direct sound so immediately that you do not become aware of its weakness.

Figure 16 shows a systematic combination of vineyard-steps and wedge-walls whereby the audience ascends in all directions but sits solely in front of the stage. The audience area is subdivided into hexagons. The thick lines signify steps of double height ($2h$), the thinner lines steps of single height (h).

Perhaps the system is better understood by the photograph of a model (see Fig. 17) which has been developed on this principle by Muller-Metge, an architect of the Neue Heimat International, for a congress building in

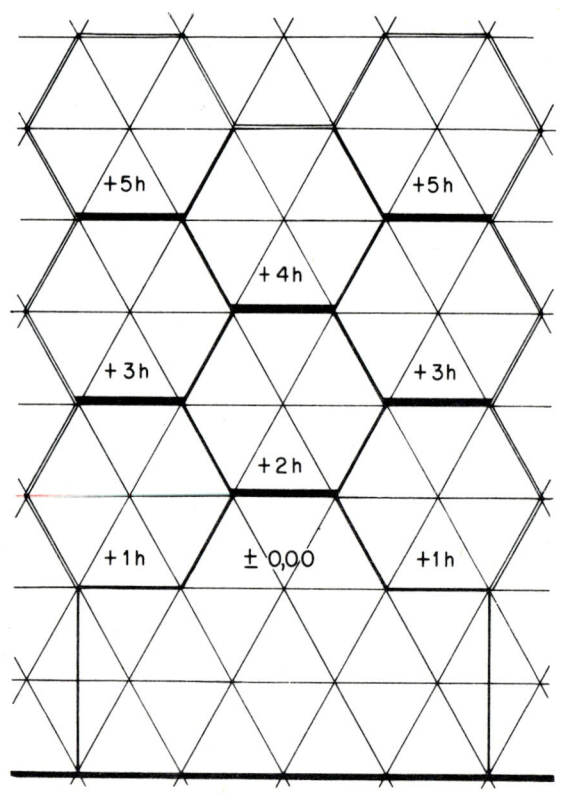

FIG. 16. Schematic diagram showing combination of vineyard-steps and wedge-walls.

Different Distributions of the Audience

Monte Carlo which should be usable for concerts too. The splitting of an audience in blocks is especially suitable for discussion at congresses. Speakers standing at the front of a block are easy to perceive, much easier than in a monotonic audience area.

FIG. 17. A model of a congress building in Monte Carlo.

Certainly this system should be combined with an electro-acoustic arrangement where the discussion microphones of one block transfer the speaker's voice to a loudspeaker hanging over this block and so avoiding discrepancy between the position of a speaker and the direction of the first soundwave to arrive.

14

The Future of the Amphitheatre

SANDY BROWN

*Sandy Brown Associates,
Conway Street, London, England.*

During 1969 and 1970, when the Rolling Stones were on tour throughout the world, they had already played in a number of open-air stadia (in the wake of the Beatles). Invariably these were sports grounds accommodating between 10 and 100 thousand spectators. Designed for visual display the type of music most readily adaptable for performance in such places was the military, or brass band variety. Simple regular beat and much visual paraphernalia—baton waving and so on. Very simple and well known melodies. This last is quite important because it is not easy to amplify the output of a marching band and the sound level output of these bands seldom reaches 100 dB in any octave band at a distance of a few metres so that, in distant parts of the ground, only snatches of melody surface above the ambient level. But if the melody is sufficiently familiar only a snatch now and then is needed to prod the memory. Conditions, in short, where any form of quality music could not survive. Now it is pretty clear that sophisticated amplification would help a static orchestra in these circumstances, as long as the orchestra could be considered as a quasi-point source. If it were not, one would be faced with acoustic delays of up to one second from one part of the arena to another. Every country and club, be it basketball, American or Association Football has its Anthem, and it is instructive here to look at at least one anthem, that of Anfield, the Liverpool Association Football supporters' ground. (This anthem has also been adopted by nearly all Association Football Supporters when their team is playing away from the home ground.) It is a composition by Richard Rogers and Oscar Hammerstein called 'You'll Never Walk Alone.' Noticeable, however, is the brevity of the excerpt from this lugubrious ballad. Supporters eschew most of the melody apart from the last 12 bars. These are sung at a funeral tempo and take 30 s to deliver, including a halting rallentando on the last 4 bars by which time the supporters may be out of time with each other by about 1·5–2·0 s. With the cunning that accompanies selection of a great team, they had selected a melody which by rallentando almost halves the tempo for the last 4 of the 12 bars and was getting quite close to the Gregorian chant (suitable for extremely reverberant situations.) An almost analogous situation, at least in that it is not one in which the time constant is critical.

The point here is that any extension beyond a 15-second melody where delays of the magnitude of a second are extant will generate insoluble problems even to well rehearsed, if unprofessional singers.

Of course this does not happen where quasi-point sources are being considered—but I should like to return to the problem which faced the Rolling Stones in the last years of the last decade. Their problem was explained to me succinctly by their road manager Ian Stewart. It is impossible to find auditoria suitable for the Stones and their fans in Britain or on the Continent of Europe. The reason is an economic one. The potential income from an album selling 100 m copies over 4 years is approximately $48 m at a price of $5 a copy, which is quite acceptable to the fans.

To match this earning potential from live concerts, at a reasonable admission fee, they would need to entertain 250 000 a week, 52 weeks a year. Huge stadia and 2000–3000 seat halls are the only possibilities; but the acoustics of stadia are poor and the fans want to *hear* the Stones.

A similar problem occurred for Johann Strauss on his first visit to the U.S.A.

He was booked to play in front of 40 000 people with a giant orchestra of about 170 players. The performance (which was received tumultuously by the audience) so unnerved him that he immediately fled the country for the safety of Vienna. His music, not unlike that by the Stones and the Beatles (who were cited as the best song-writers since Franz Schubert by William Mann of *The Times*) requires reasonable acoustic conditions without undue delays or reverberation.

Reasonably sized halls were the better alternative using listening conditions as an absolute criterion but they demanded enormous seat prices which the Stones had no wish to impose. Even if (as had been considered) the Stones hired the halls themselves and ran their own concerts in them charging low seat prices, a black market would have sprung up in ticket sales. The other alternative, sports stadia and the like, would, and did solve the ticket cost problem but in spite of using sophisticated portable amplification, listening conditions for most of the auditors were predictably abysmal. Furthermore climatic conditions in Northern Europe and more than half of the United States ruled out performances in the open air for six months in the year.

Now it could be argued that the Stones, the Beatles and a handful of similar popular groups are extraordinary cases but record sales do not altogether support this argument. There have always been, in living memory, musical attractions which draw vast audiences. In the '30s and '40s it was the day of the big bands—Glenn Miller, Artie Shaw, Benny Goodman, Jimmy & Tommy Dorsey, later Bill Haley and other rock n' roll artistes in the '50s, the Beatles, Stones and Cream in the '60s and Emerson, Lake and Palmer and numerous other groups today.

E.L.P. are still facing the same problem posed by Ian Stewart. They feel that they can only make 4 live performances in the U.K. this year in order to minimise it. To this problem there would seem, at the moment, to be no answer except to sit at home listening to very popular artistes on your stereo—or quadrophonic—Hi-Fi.

So far we have considered only a situation which could be described as market-controlled patronage. The market, demonstrably, is controlled by sales of recordings. 'The Charts,' 'The Top Ten' and so on refer exclusively to indices derived from numerical sales or recordings. It is clearly more profitable to press and sell 10m copies of one recording than 1m copies each of 10 recordings—or yet 10 copies each of 1m recordings. So economic incentives abound to jump on any bandwaggon which will further this aim. Huge investments are therefore loaded in a direction which will ensure that large numbers of a small selection of recordings are marketed, so that any performances reliant on market-controlled patronage are at a serious disadvantage if unable to prove in advance that heavy investment is a reasonable risk. Latter day Mozarts will continue to suffer penury and as the world has had some hundreds of years to consider this problem without tangible success I hesitate to enter the field.

But let us now consider state or national patronage, which has taken the place of private patronage in Europe and Trustee patronage which, together with corporation patronage, has undertaken this role in the U.S.A.

No symphony orchestra in the world derives its revenue solely from recordings and ticket sales. Subsidies vary, but in Northern Europe the subsidy could be as high as 90% of revenue. It is in Copenhagen (for Opera) and in Recklinghausen for the main multi-purpose hall. The Royal Opera House, Covent Garden, receives an annual subsidy from the Arts Council of Great Britain of nearly $4·5m—almost 10% of the Arts Council's annual budget. For elaborate operatic performances every member of the audience is virtually handed $50 to go in. I see nothing wrong with that: the alternative at present would be to disband the highest standards of operatic and symphonic performances. But this is a holding operation. Furthermore, subsidies usually comprise a non-indexed system of dispensing favours. They are not linked to any inflationary spiral except through the donors who themselves are under pressure from the spiral while they make their decisions.

Is there another possible alternative? Well, on the face of it, there is not. You cannot have 20th Century musicians and artistes entertaining a 20th Century audience in an 18th Century hall at 20th Century earnings and ticket prices without a major change in structure. At present the structure survives by patronage on a financial scale unknown in bygone times. But apart from the drawbacks of corporate patronage (timidity, playing for safety and rather more insidiously, élitism), there is also an echo of the Rolling Stones' problem, that the most popular performances are vastly oversubscribed. Queues for some Covent Garden performances start the previous night. Where 20th Century audiences are concerned we may as well recognise that although no electronic methods exactly substitute for the real performance audiences are now thoroughly used to electronic sound reproduction. Moreover we should consider the quite vast differences which have entered music in the last 50–100 years. A performance of a Beethoven symphony as Beethoven might have heard it would now be totally unacceptable. Strings have played with a vibrato for about 50 years: they did not before that.

Some of the notes have been changed (Beethoven's trumpets had no valves) the timbre of most of the instruments has changed (an Albert system clarinet has fewer holes and therefore a different timbre to a Boehm system instrument). Soon we will have the logical Bassoon, of which a prototype is already available. What about the logical auditorium? I thought it about time that someone had a stab at it so I present the Phonodrome.

This seats 20 000 people which is nearly 10 times as many as Covent Garden and in economic terms would solve a difficult problem for those blessed (and cursed) by public subsidy and an impossible problem for the Rolling Stones and their offspring. And if I say, as an assertion, that there is nothing so intractable as human attitudes, I do so in the knowledge that assertion itself is an example of this intractability. Nevertheless it is an assertion with which few would disagree. Attitudes, including a reluctance to change, must be an integral part of the human survival mechanism if Darwin is right. They manifest themselves in totem, taboo and religion. They are not often amenable to change by force and even less to change by persuasion. But the subsidy alternative works poorly, inequitably, and is seriously endangered by inflation so I propose that we will have to alter public taste. I do not think I have to underline the difficulties here. If we forget about real acoustics altogether, the design problem of large arenas is quite straightforward. Plenty have been designed and built during the history of man.

But as we expect the 20 000 who visit the Phonodrome to be listeners more than watchers, we must replace the 'real' acoustics with 'artificial' acoustics. This is already what everyone does at home and, as I have said, they are used to it. They certainly are not used to it in operatic auditoria, but almost everywhere else—in the concert hall, the cinema, and so on— electronic aids are normal and acceptable. I do not discount the fact that something will be lost but it is fair to add that there is also a great deal to be gained.

Figures 1(a) and 1(b) show a plan of the Phonodrome in an open position, and it can be seen that there is an off-centre arena in the middle and that it is somewhat fragmented. One problem of seating 20 000 people looking at a group singing is that, some of them will be facing the wrong way or,

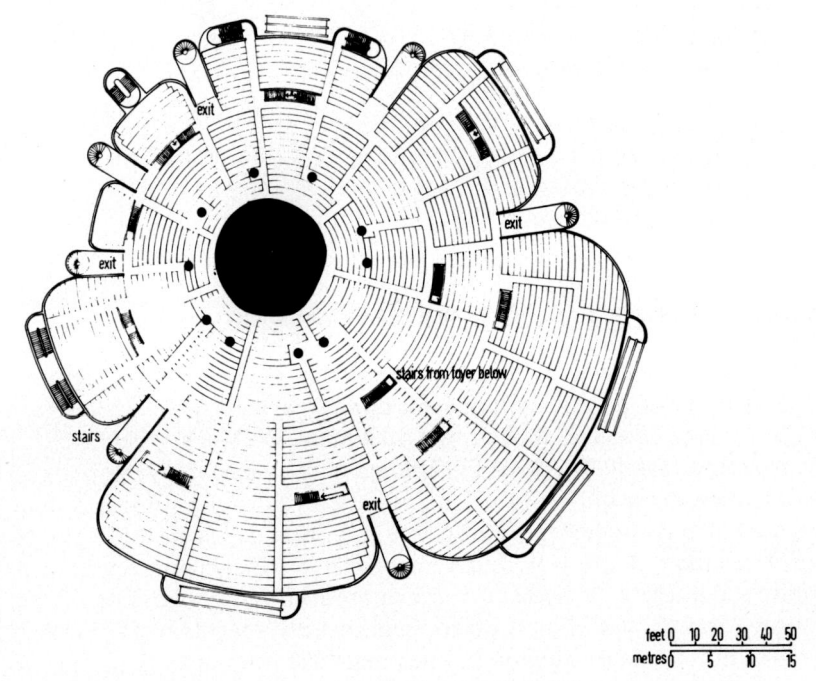

FIG. 1(a). Phonodrome—main auditoria combined with maximum seating.

rather, the ceilings may be facing the wrong way, and to put them back the right way one would have to have visual as well as audio aids. These could be suspended high up so that the audiences could look at the blown-up picture of what they wanted. This would also aid people sitting in the far rear because, visually they are quite a long distance away.

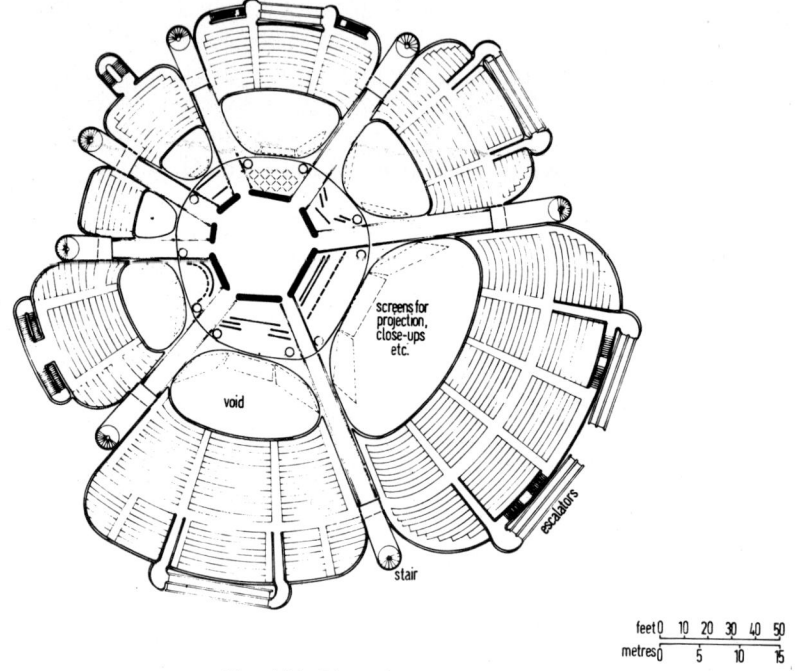

Fig. 1(b). Phonodrome—upper levels.

The segmented condition can be seen now in Fig. 2 where the segments are divided by large doors which come down from the roof. The doors have already been invented, in fact, by Dr. Schultz. One such system which I engineered myself for the ATV Studios in Wembley, provides a 70 dB separation, therefore with the kind of air gap shown here which is more than 3 m, one would expect to achieve from mass law, an average insulation of nearly 80 dB. It is clearly impossible in the adjoining segments to have a pop group adjacent to a drama production, because 80 dB is not nearly enough, but in most other circumstances it would work. Furthermore, in the smaller auditoria, which seats 900 it is perfectly possible to have an almost purely acoustic situation, but as the whole place would have

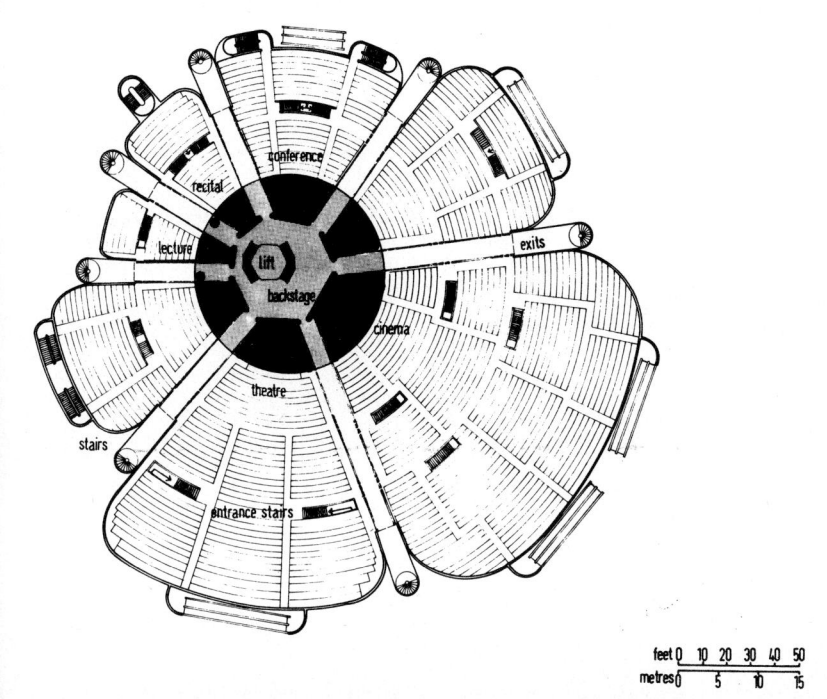

Fig. 2. Phonodrome—auditoria set up for independent use.

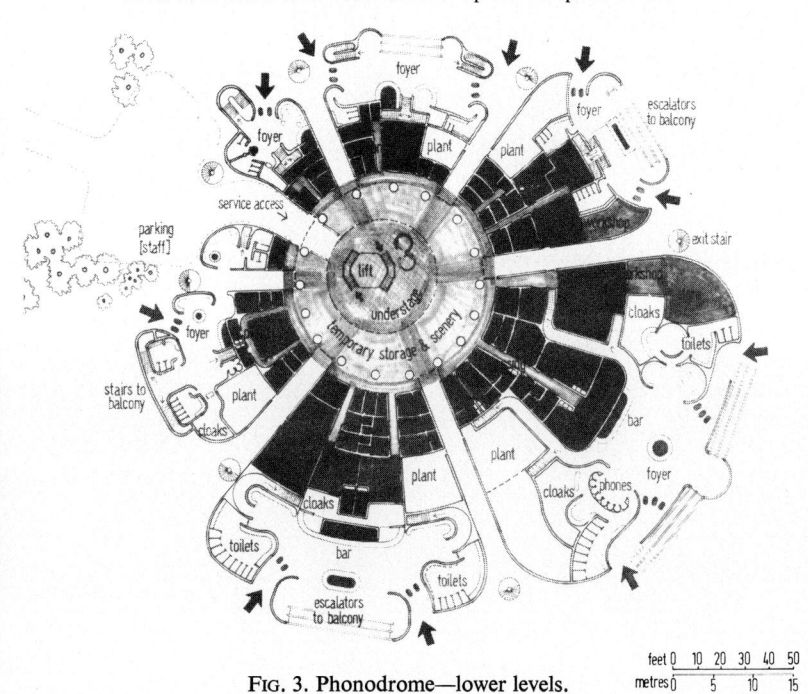

Fig. 3. Phonodrome—lower levels.

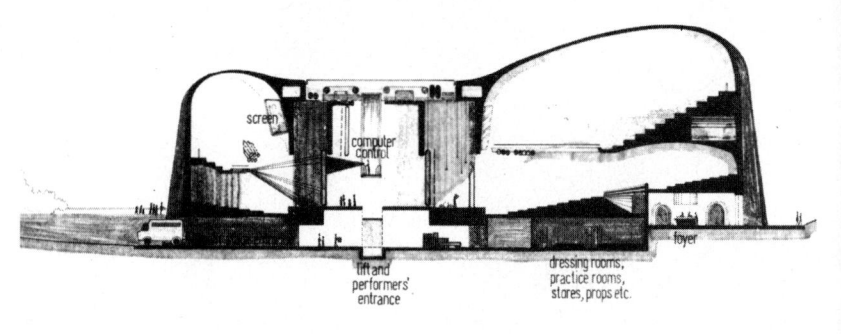

FIG. 4. Phonodrome—section through large and small auditoria.

to be extremely dead, it would never in any circumstances be possible to have a completely acoustic performance.

Finally I should like to end by quickly referring to the remainder of the figures. Figure 3 shows the service area underneath the auditorium and the various entrances for maintenance. Figure 4 is a section through the Phonodrome showing the large and small auditoria.

15

Assisted Resonance

PETER H. PARKIN

Building Research Establishment, Garston, Watford, England.

I am confining my paper almost entirely to reverberation time, which I am old-fashioned enough to think is still a very important factor in the design of auditoria. Of course it is not the only thing—it never has been—but it is important. I do not know quite why it is so important. Obviously you hear the music decaying, and this bridges the gap between the various notes. In extremely reverberant conditions like cathedrals it is obviously very important for the type of music that was written for performance there. I think perhaps more important is that, in the sort of auditoria we are mainly concerned with here (that is, theatres and concert halls) it is a measure of the ratio of the direct to the reverberant sound energy. I am absolutely convinced that sound coming from all round you, that is the reverberant sound, is important for 'envelopment,' or any of the other terms that are used.

This leads me on to the criteria that I would use for judging any artificial reverberation system. The first is the obvious one, the reverberation time measured in the usual way—the first 10 dB, or the first 5 dB to 35 dB or whatever your preference. The next thing you measure is the power increase in the auditorium due to the artificial system. By the term 'artificial' at the moment, I am including all electrical and mechanical alterations. Any change in the reverberation time should correspond to an increase in the power. (This to a certain extent takes care of Dr. Jordan's worry about the early part of the decay; if in fact it does not start from the top then you will not get the same power increase that you should do.) The last criterion is very obvious; it is that it should sound satisfactory and not be distorted.

I want to give a very brief review of what systems there are for altering the reverberation time of auditoria either permanently or from moment to moment. I should say I have not heard any of these systems except the assisted resonance one, and I shall only be going on what other people have told me. In previous papers the mechanical systems, by which I mean altering panels and ceilings, have been described. I am a little surprised at the figures given for the changes; I know if a hall is designed with this sort of change in mind then you can do much more. I am surprised you can get so much change at low frequencies. I am surprised it is not a lot more expensive than was quoted; perhaps in America they are very much better at it than we are in this country. I shall not discuss the mechanical changes—they have been adequately covered. Figure 1 shows the Royal Festival Hall and the wooden panels on its side wall. If these were made changeable then the extreme limits one could conceive as practicable would be a 6% change in reverberation time at 125 Hz, and at 500 Hz a 19% change. This, of course, is because the audience constitutes much of the absorption of 500 Hz, and the other surfaces at 125 Hz.

Fig. 1.

One electrical system is known as ambiophony. Professor Kuttruff has given a description of his system (see paper 12). There are other examples of that type of installation although almost certainly not as well done as his was. I think there is perhaps an artificial line to be drawn here. He mentioned close microphone working and how that sort of system will only work if you can get the microphones close to the source. The B.B.C. did some work some years ago and in order to get their system to work in a television studio no musical instruments could be placed more than two metres away from the microphone. This is a question of power which I referred to earlier. It is difficult to get enough power into the auditorium from that sort of system if you have got remote microphones because you run into feedback troubles. It is perfectly satisfactory for broadcasts and gramophone recordings, but I do not think that, in a concert hall, it would normally be acceptable to have these close microphones working, certainly not in the Royal Festival Hall. Franssen published a theoretical paper in 1968 on a regenerator/reverberation system using a large number (50—in order to double the R.T.) of loudspeakers and microphones all feeding back over a broad range, that in theory look extremely good. The Russians have just published a paper saying that it cannot possibly work but I believe Phillips are working on it; it will be interesting to know upon completion how successful it is.

Guelke and Broadhurst in 1971 described another system relying on direct feedback. It relies to a certain extent on directional loudspeakers and directional microphones. Perhaps that is a limitation although it is obviously working well where it is. Jones and Foweather have been working for some time in Manchester. In their particular hall they have a reverberant space above the ceiling. The energy from the main auditorium flows through holes in the ceiling up to that reverberant space. Microphones in the reverberant space feed that back to loudspeakers in the auditorium. Now that at first sight, to my mind is very little different from ambiophony and should suffer the same limitations, but the Manchester installation does seem to be much more successful and all that matters is what it sounds like. I am not quite convinced by their explanations of how it works, but the main thing is it does. These I think are the main systems one might call currently contending for this market, so I can now get onto assisted resonance—the main subject of this talk.

First, a very brief history of the acoustics of the Festival Hall. The reverberation time intended originally by Hope Bagenal was 2·2 s. During the design stage it was obvious we were not going to get that without a very high volume. This is an example of all the conflicting things the architect has to cope with referred to by Sir Robert. Another example was that the orchestras wanted a much larger hall than was, in fact, built; but Mr. Bagenal wanted it to be much smaller. The design compromise in the end was 1·7 and it came out when the Hall was finished at 1·5 s. In spite of this the Hall was very popular at the time and still is very popular with the concert going public. But there have always been complaints about the 'dryness' particularly from performers and in 1964 when the Greater London Council (who own the hall) were going to close it for redecoration, (the first time since 1951), we considered what could be done to reduce the 'dryness' which we assumed, perhaps wrongly, meant increasing the reverberation time. There were many things that we could conceivably have done but what we did do was put in the assisted resonance system. The general principle is again based on regenerated reverberation (see Fig. 2). You put a microphone and a loudspeaker at an anti-node of one peak in the response to the room. You join them with an amplifier, you adjust the

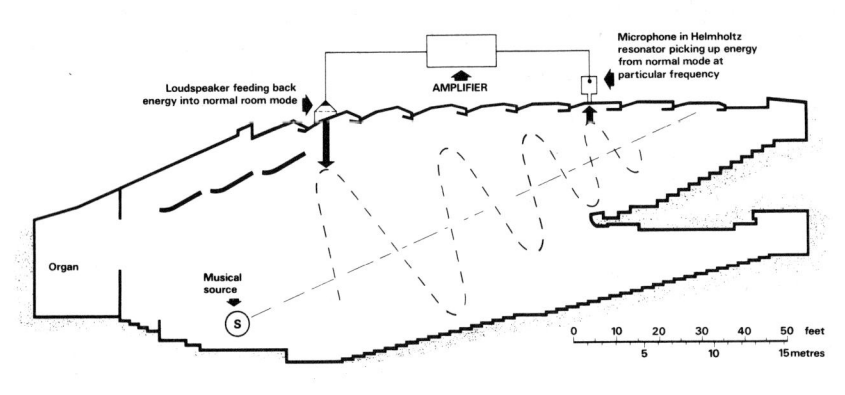

FIG. 2.

phases so the signal is in phase around the loop. If you turn the gain up enough it all feeds back. If you turn the gain down for example to 6 dB below feedback then you double reverberation time at that one frequency. This has to be done at a lot of frequencies, and in the Festival Hall there are now 172 such 'channels', covering the range from about 60 Hz up to about 700 Hz. I should say that the original idea was rather like the way Franssen is now doing it, that is, by positioning the microphones and loudspeakers in space without filtering; but we soon found that 5 or 6 such channels found some frequency of their own which they all liked responding to and this was hopeless. What we should have done was, perhaps, to have gone on like Franssen has done—if you get enough channels (say 50) then they start cancelling out and you're back where you were, but with double the reverberation time, which is, of course, the object. In our system the filtering to stop channel interaction could have been electrical, but in practice it was much easier to put the microphone into a Helmholtz resonator, which is cheap, reliable and lifts the electrical signal 30 dB—so there is no trouble with electrical noise changing frequency with temperature of the hall. The whole unit is fairly foolproof. One small point, Helmholtz was apparently asked by the King of Bavaria if his work on resonators could have any application to making acoustics of public buildings better and Helmholtz said he could not imagine any possible way they would be of help.

Figure 3 is a typical transmission irregularity which you are all familiar with. This particular loudspeaker and microphone position would be suitable for, say, 455 Hz. Each of these peaks will have its own reverberation

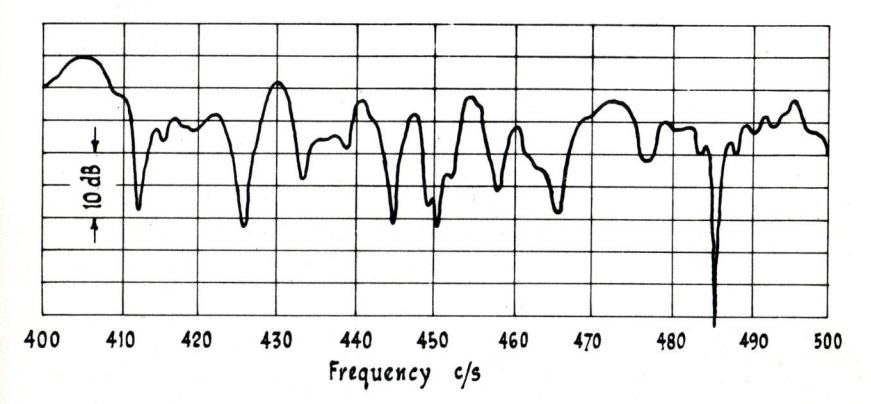

FIG. 3.

time, varying considerably in shape, on top of the variation in sound pressure level, which you can see on the graph, and it is almost an art to decide which position on that sort of graph you would choose to put your microphone. Coming back to this question of power, you do not have as many normal modes excited by this system as in real life. You do not know how many you do need. Nature probably provides far too many normal modes as she does most other things. The point I am making is that if you choose one of those peaks which has a fairly low pressure, and also choose one which has a shortish natural reverberation time, then you have to increase the gain of your channel by a few dB more to bring it up to 2 s (say) than you would if you chose a position with a relatively long natural reverberation time. Thus the choice of a low shortish peak is a way of putting in more power to the system than you would get from a position chosen at the top of a peak.

174   *Auditorium Acoustics*

I mentioned resonators as the most convenient way of filtering. Figure 4 shows one of the various sizes used with a microphone inside. Tuning is achieved by moving the back plate in and out. There are some of these resonators above the ceiling in the Festival Hall, as can be seen in Fig. 5. In the original design of the Festival Hall ceiling some 1200 five centimetre-diameter holes for resonators were left. This was based on the experience, a year or two before the Hall was designed, of Drs Jordan & Ingeslev in

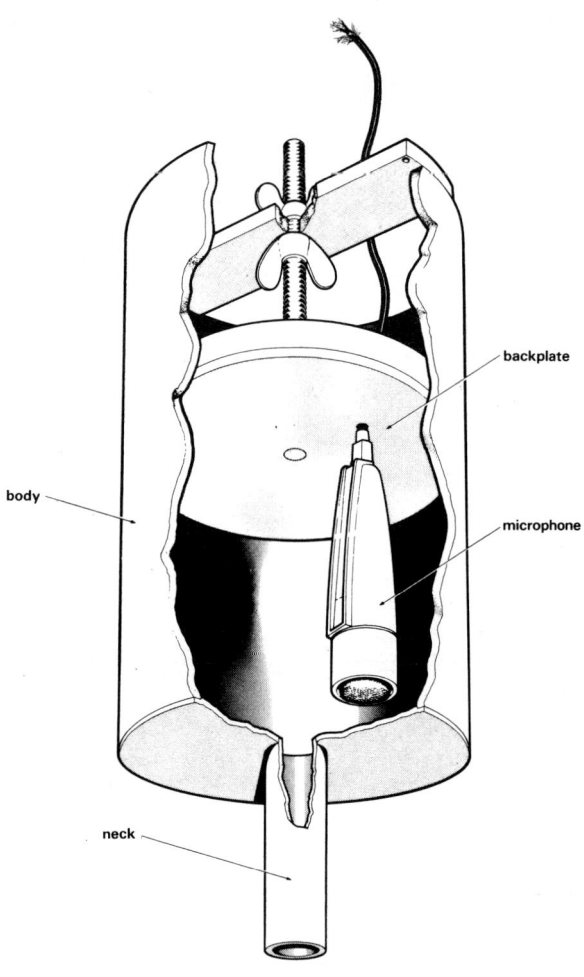

FIG. 4.

Assisted Resonance

Copenhagen where they had had trouble with low frequency reverberation—they put resonators in the ceiling to absorb it. These holes were left in the ceiling of the Festival Hall in case of similar troubles. Of course, they were not used; but then they came in very useful 15 years later for exactly the opposite purpose. The resonators for the assisted resonance system have their necks sticking through these holes in the ceiling so that

Fig. 5.

the bottom of the neck is flush with the underside of the ceiling. Figure 6 shows some of the loudspeakers. There are three rows then in the lighting troughs; they cannot be seen except from the back row of the choir. One small point, below 100 Hz it is difficult to get enough power out of reasonable size loudspeakers, so below 100 Hz the loudspeakers are on quarter wave tubes, which gives you about 8 dB gain in efficiency in power output (see Fig. 7). Since each loudspeaker is only operating on one frequency, there are no complications about quality of loudspeakers, so you can use the cheapest loudspeakers that money can buy. The main problem about the assisted resonance system has been the variability of the reverberation time of the individual channels (see Fig. 8). This is only a reflection of the variability of the natural reverberation times and we do not know why they vary, it is probably something to do with temperature gradients.

Fig. 6.

Fig. 7.

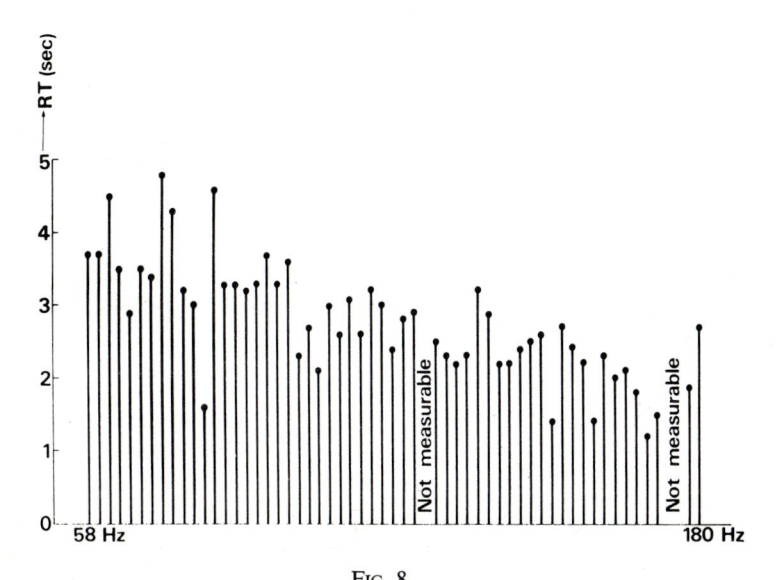

Fig. 8.

Fig. 9.

Now, coming up to date (see Fig. 9) is the Central Hall of York University, which is an attempt to install a much cheaper system so as to make it economic for a multi-purpose auditorium. This Hall was designed primarily for speech, but the architect did have assisted resonance in mind as a possible use to make it suitable for music. The main economy was to cut down the number of channels from the 172 used in the Festival Hall, to the 72 used in York. Figure 10 shows the channel spacing at the bottom and top ends of the frequency scale—and the channels at York are spaced out very much more widely than in the Festival Hall. One cannot be certain, but almost certainly we could manage with a yet wider spacing at the low frequencies and a narrower spacing at high. (Incidentally if the whole of the ceiling at York, which is woodwool, was plastered one would only get a 0·1 second increase at 125 Hz.) There are 3 settings at York. I think

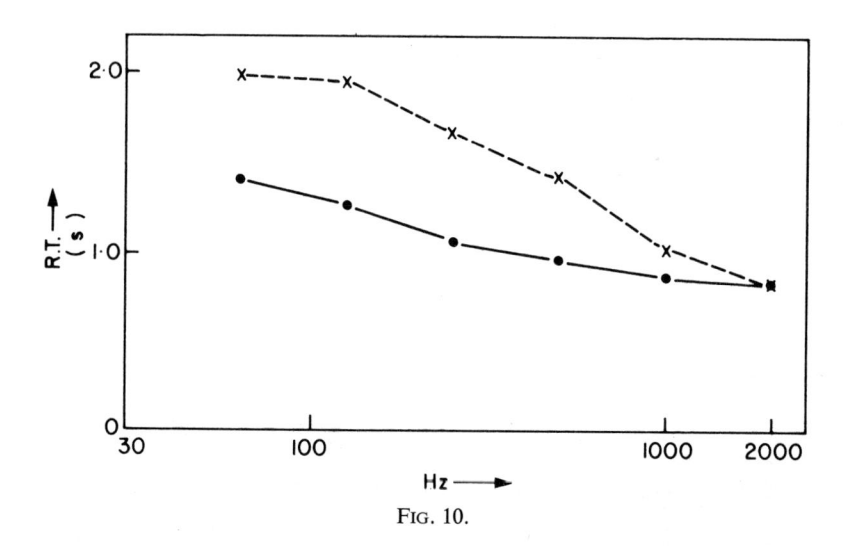

FIG. 10.

I can sum up the results to my mind at York: first, it is used for speech at one of the lower settings. They much prefer it on for speech, that is conferences, lectures, etc. On the higher setting for music, the hall is much better for small groups—chamber orchestra, soloists and so on. It is not so much of an improvement for larger orchestras and this is largely due to the shape of the hall. There is one long side with five sides facing it. It is not a very big hall—1200 seats. To get the university orchestra in, which is about 90 strong, you have to take out practically all the seats on the ground floor, this means that about 75% of the audience is in the direct field, so one cannot expect the reverberation to have much effect.

I think what we have learnt from the Festival Hall and the York experience is that you can do what you like if you have enough channels but if we talk about some economically reasonable limit of 100 channels, then you can get 100% increase at the lower end of the frequency scale, and at least a 60% increase at the mid-frequencies, that is up to about 1500/2000 Hz—above that the system is not economic. If we take a pessimistic 60% increase at mid-frequencies, the effectiveness obviously depends on where you start—if you start at, say, 1·3–1·4 s you can get up to 2·0–2·1 s—which is probably enough.

16

Problems of Sound Reflection in Rooms

ZYUN-ITI MAEKAWA

*Faculty of Engineering,
Kobe University,
Rokko Nada Kobe 657,
Japan.*

1. INTRODUCTION

Sound reflections in a room naturally play a very important role in the subjective evaluation of acoustic performance of the room, and favourable reflections should be provided by the walls and ceilings.[1] Obviously, the judgement on whether the reflected sound is favourable or not should be made psycho-acoustically. Unfortunately separate investigations have been made in the fields of psychological acoustics and physical sound transmission in rooms, so that the connection between them in the acoustical design of the room is obscure. One of the reasons why the connection is not satisfactory is the difficulty in measuring the complex reflection coefficient of a boundary as a function of the frequency with a fixed angle of incidence.

Concerning the sound reflection from even boundary surfaces, since the first theoretical investigation was done by Green[2] in the early 19th century, more than a hundred years have been needed to measure the complex reflection coefficient, or the acoustic impedance, as a function of the angle of incidence,[3-6] and the 'Interference Pattern Method' has been developed as a standard method for both even and uneven boundary surfaces.[7]

There are many examples of rooms having reflectors, and so-called 'clouds' have been often used in auditoria.[8,9] Furthermore, we can never forget the unfortunate event of 'clouds' in the Philharmonic Hall in New York. The sound reflection from a panel or a panel array has been studied by several authors[10-14] who however studied only the physical aspect in both theoretical and experimental investigations. Naturally, we should continue to find new methods to design sound reflections and provide good acoustics for the audience. A very suitable way is to simulate reflected sound in an anechoic room to carry out psychoacoustical experiments. Classically, an analogue simulation method was used for this purpose.[15,16] Now, we can use a digital computer as a useful tool[17] to study the reflection transfer-functions.[6] This report reviews recent work in this area of study.[7,18,26] A sound reflecting wall in a room can be regarded as a linear system. Let the complex frequency response of the wall be $W(\omega)$, with an inverse Fourier transform of $w(t)$. They should be called the reflection transfer-function and the impulse response of the wall respectively. They are also functions of the positions of source and receiver $(\mathfrak{r}|\mathfrak{r}_o), \mathfrak{r} = (r, \theta, \psi)$, corresponding to the Green function of the sound field. In Fig. 1, let the

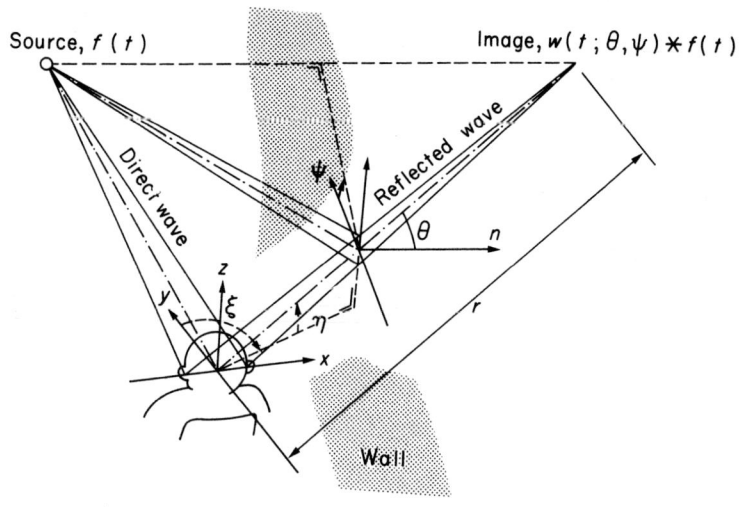

FIG. 1. Positions of listener and reflecting wall for the simulation of reflected sound.

source function be $f(t)$ and the reflection impulse response of the wall be $w(t,\mathfrak{r}|\mathfrak{r}_o)$, then a sound reflected by the wall, $g(t, \mathfrak{r}|\mathfrak{r}_o)$, is simply given by

$$g(t,\mathfrak{r}|\mathfrak{r}_o) = w(t,\mathfrak{r}|\mathfrak{r}_o) * f(t), \tag{1}$$

where the symbol * shows a convolution integral. The reflected sound and its set might be simulated by the aid of a digital computer with A–D and D–A converters.

2. TRANSFER FUNCTIONS AND IMPULSE RESPONSES OF BOUNDARIES

In order to simulate a reflected sound from a boundary, it is not sufficient to know the sound absorption coefficient by the reverberation-chamber method or the normal acoustic impedance of the boundary. We must find its sound-reflection transfer-function, *i.e.* the frequency characteristics of the complex sound pressure reflection coefficients, or acoustic impedances for an arbitrary incident angle (θ, ψ) as shown in Fig. 1.

Examples of transfer-functions measured by the 'Interference Pattern Method' are shown in Figs. 2 and 3. The function of these materials becomes $W(\omega, \theta)$ since they do not depend on the azimuth angle ψ.

The impulse responses of sound reflections from these boundaries

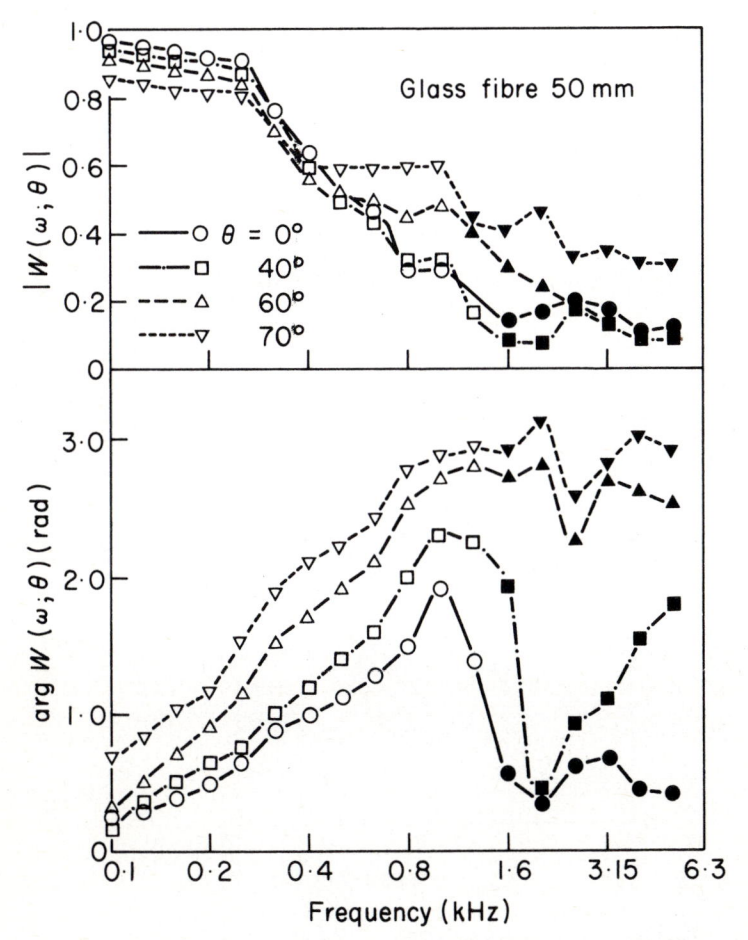

FIG. 2. Measured transfer functions $W(\omega; \theta)$ of 50mm-thick glass-fibre.

naturally can be obtained by inverse Fourier transforms of the transfer-functions by the aid of a computer, as shown in Figs. 4 and 5. Their parameters are the angle of incidence. These values should be used in calculation of Eqn. (1). In order to obtain an impulse response at a boundary, there are other methods using correlation techniques,[19-21] using a single pulse [22,23] or an adequate digital filter,[24] though they must be examined with regard to their accuracies.

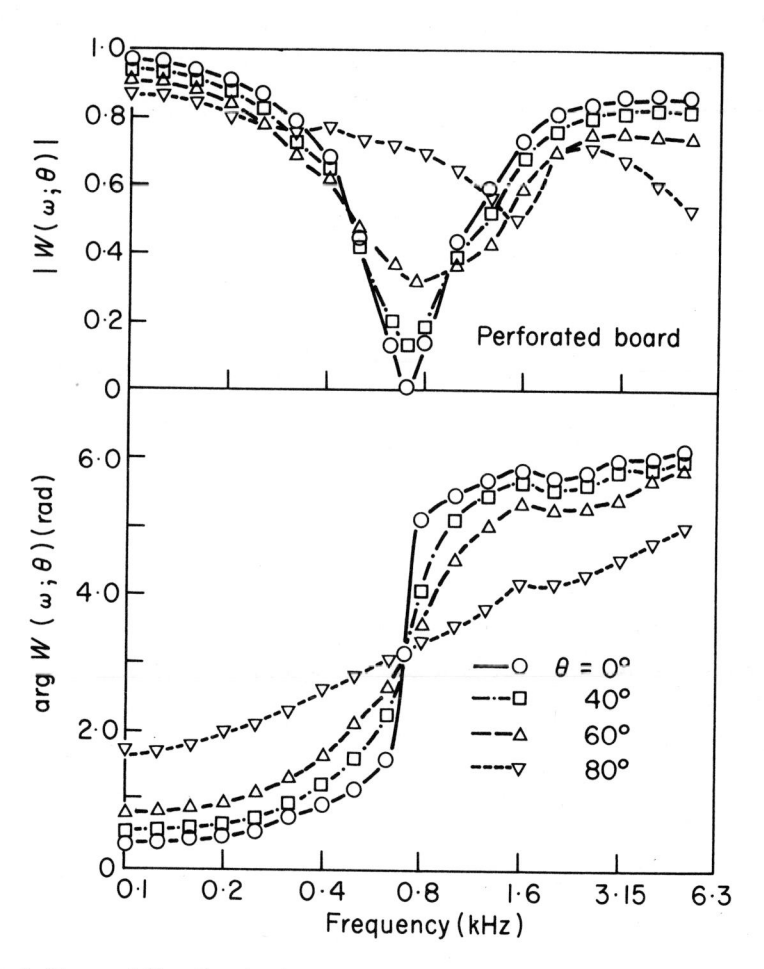

FIG. 3. Measured ($\theta = 0$) and calculated ($\theta \neq 0$) transfer functions $W(\omega; \theta)$ of a perforated board, after Ingard.

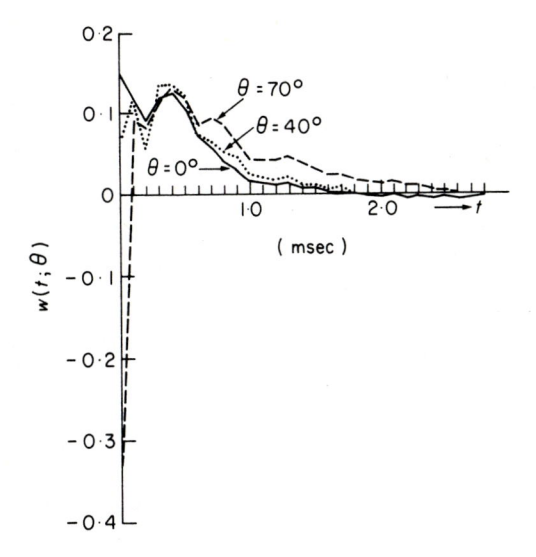

FIG. 4. Impulse response of glass-fibre, at a fixed angle of incidence.

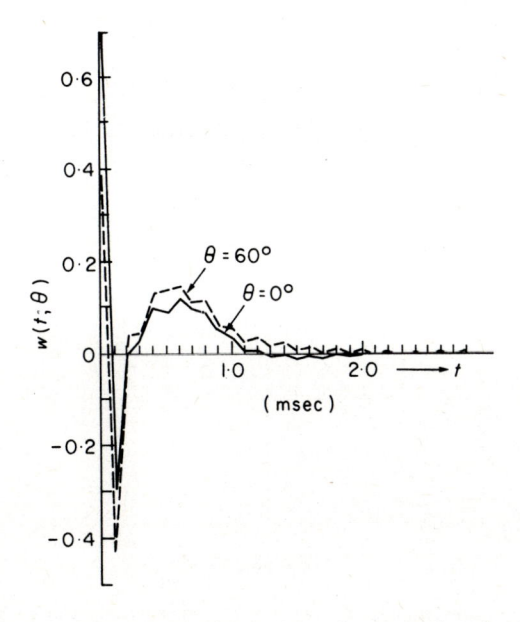

FIG. 5. Impulse response of the perforated board, at a fixed angle of incidence.

3. SOUND REFLECTIONS FROM UNEVEN SURFACES

In room acoustics, one of the most difficult problems is to deal with irregular shapes or uneven surfaces. Here, let us show some examples of the availability of the sound reflection transfer-functions for uneven surfaces. Figure 6 shows a structure of uneven sound absorbing surfaces whose com-

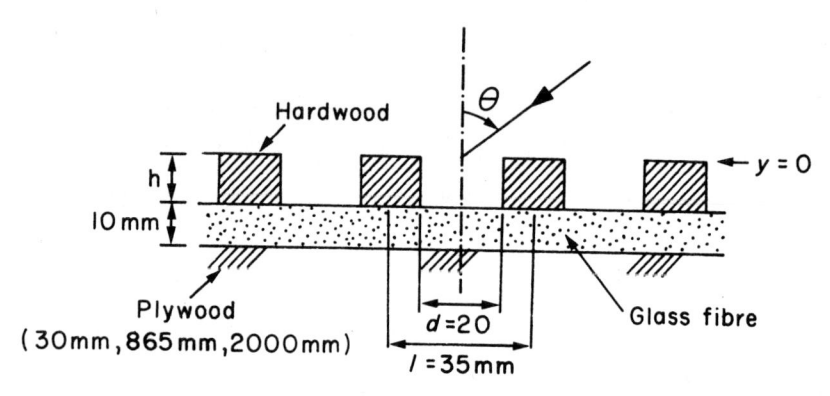

FIG. 6. Periodically uneven surface of rectangular profile.

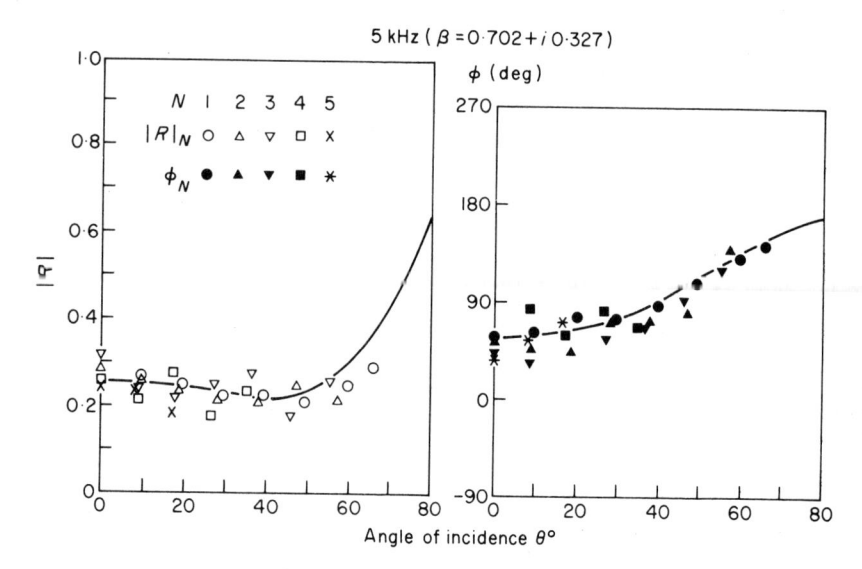

FIG. 7. Measured and calculated reflection coefficients of 10mm-thick glass-fibre, assuming 'locally reactive' surface.

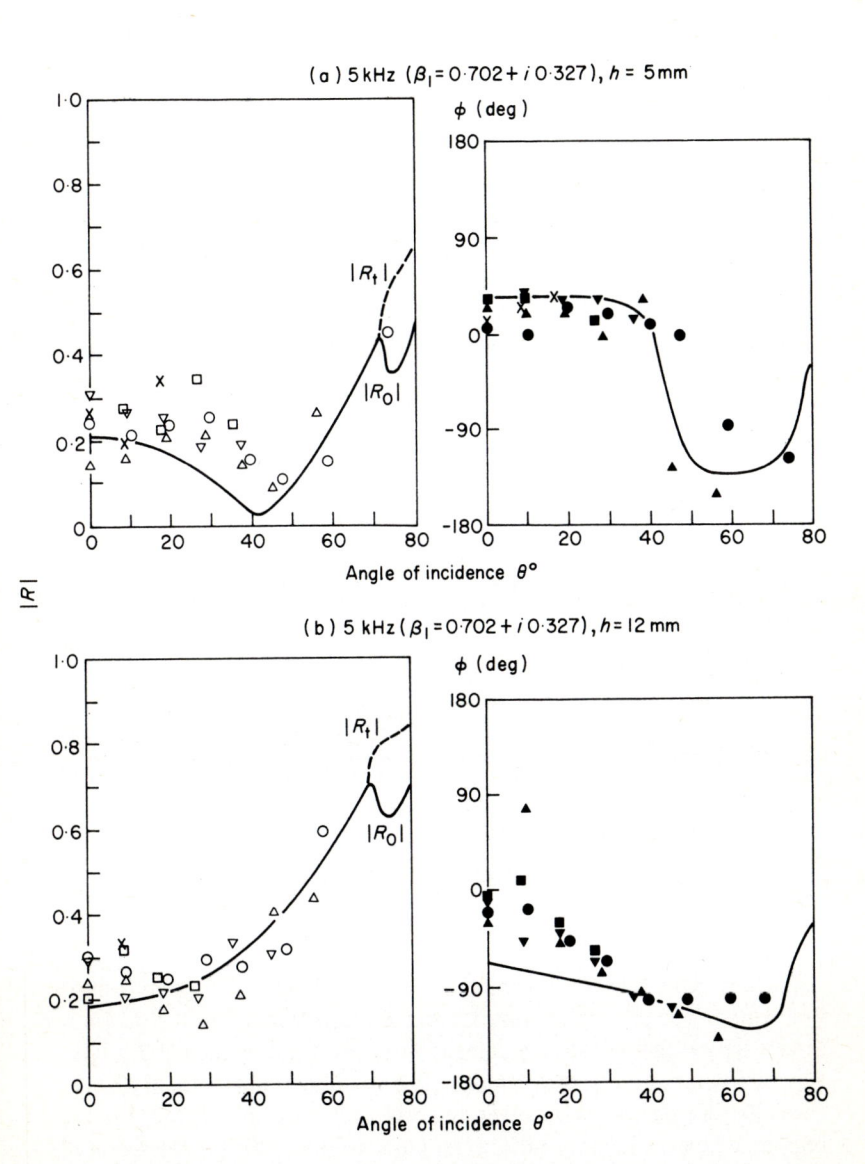

FIG. 8. Measured and calculated reflection coefficients of the uneven surface of periodic rectangular profile, as a function of angle of incidence.

plex reflection coefficients may be measured and calculated. At first, the sound reflection coefficients of bare glass-fibre, without hard-wood ribs were measured by the Interference Pattern Method[6] and calculated with the assumption of the existence of a locally reactive surface, as shown in Fig. 7. Then, with wooden ribs periodically laid on its surface, as shown in Fig. 6, the complex sound reflection coefficients were measured by the same method, and compared with the theoretical values calculated by Bruijn's theory[25] as functions of the angle of incidence at 5 kHz, as shown in Fig. 8. (Figure 8 (A) and (B) show the results when the height of the wooden ribs (h) in Fig. 6, are 5 mm and 12 mm respectively. In these figures, both $|R_0|$ and ϕ_0 show the theoretical values only for a component of specular reflection, and $|R_t|$ shows the total reflection coefficient including scattered waves.)

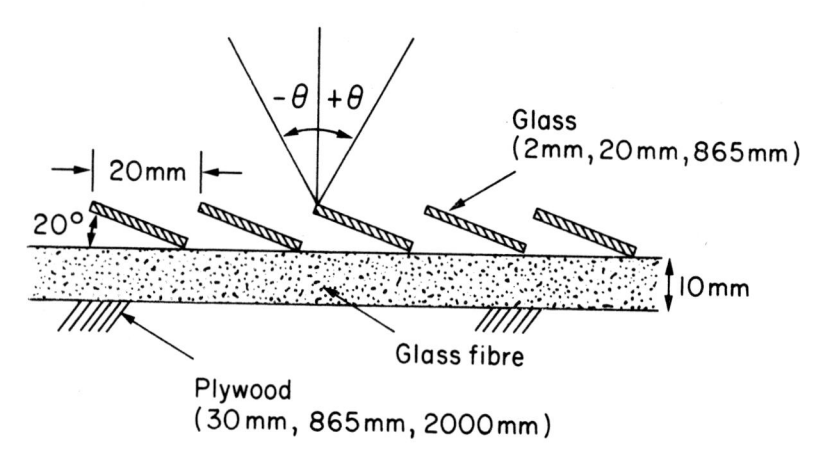

FIG. 9. The louvre array obliquely attached to the glass-fibre.

In these cases, the components of specular reflection are prominent, and the values measured by the Interference Pattern Method in the region near the structures show good agreement with theoretical values of R_0, both in absolute values and in their phases. The usefulness of the theory has been verified, though the experiment is successful only in the low frequency region. When the height of the rectangular wooden ribs (h) changes, reflection coefficients at oblique incidence are widely varied. Comparing with the values of the bare glass-fibre in Fig. 7, the values of reflection coefficient for these structures are unexpectedly reduced by the wooden ribs laid periodically on the surface of the glass-fibre, especially in Fig. 8(A).

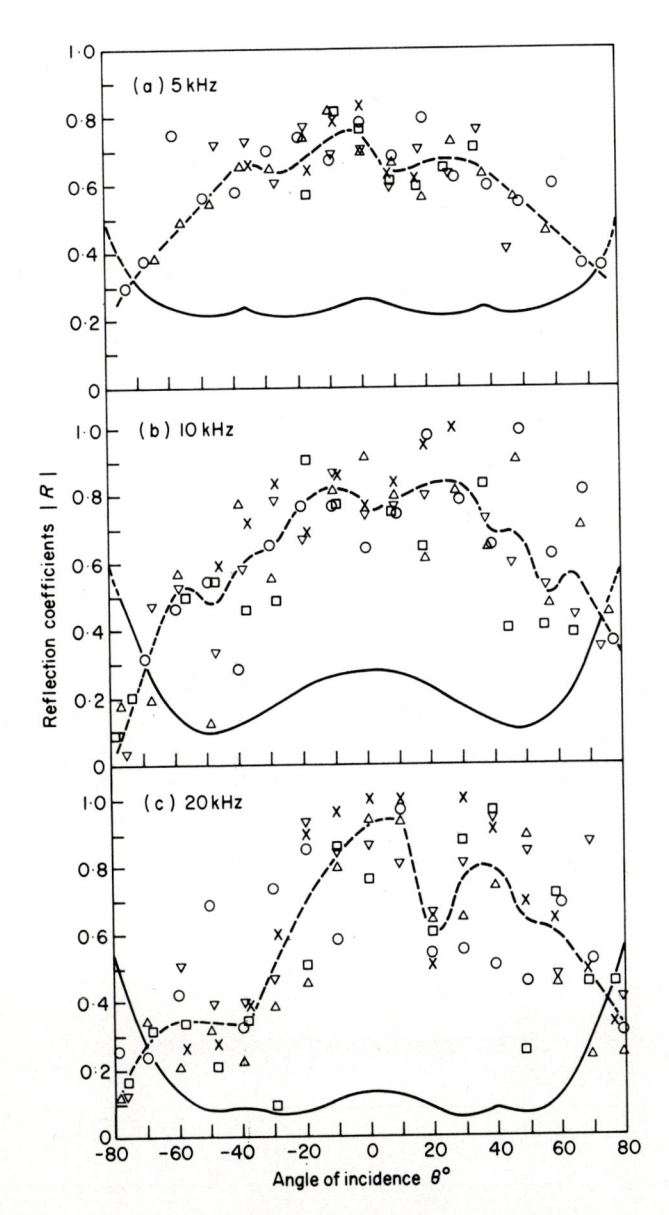

FIG. 10. Measured reflection coefficients of the louvre array obliquely attached to the glass-fibre, as a function of the angle of incidence, (dotted lines). Solid lines show the measured values of bare glass-fibre as in Fig. 7.

Another structure used is shown in Fig. 9. The louvres covering the surface of the glass-fibre were made of glass plates. This type of structure is very interesting since we can expect directional absorption, though this has not been analysed theoretically yet. The results measured by the Interference Pattern Method are shown in Figs. 10(a), 10(b) and 10(c) at the frequencies of 5, 10 and 20 kHz respectively. The asymmetry with the sign of the incident angle at 5 KHz is not so obvious, but it is clear at above 10 KHz. It may be seen that the reflection coefficients increase in the region of positive values of the angle of incidence. Furthermore, it is a very interesting fact that the reflection coefficients of this structure in the region $|\theta|>70°$ are smaller than those of the bare glass-fibre (dotted lines, as shown in Fig. 10) at all frequencies.

4. SIMULATION OF REFLECTED SOUND AND SUBJECTIVE TESTS

Using the measured or calculated transfer-function of the walls, reflected sounds were simulated by Eqn. (1) by the aid of a computer, and some subjective tests were performed.

4.1 Timbre of Reflected Sound

At first, eleven subjects participated in paired comparison tests, based on the listeners' ability to distinguish the difference of timbre. A reverberation free speech-signal (duration 1·2 s) was convoluted with the impulse responses of walls, of glass-fibre and a perforated board, for various incident angles as shown in Figs. 4 and 5. The simulated sounds $f(t)* w(t, \theta_1)$ and $f(t)* w(t, \theta_2)$ were radiated from a loudspeaker in an anechoic chamber and the subjects were requested to judge whether they recognised the difference of timbre or not between two signals (using two ears mono-

TABLE 1
Percentage judging timbre as different

(a) For glass-fibre			(b) For the perforated board		
$\dfrac{\theta_2}{\theta_1}$	40°	70°	$\dfrac{\theta_2}{\theta_1}$	40°	60°
0°	9	86	0°	5	81
40°	–	86	40°	–	55

phonically). It is clear from the results shown in Table 1 that more than 80% of subjects can recognise the timbre as different when comparing an angle of incidence $\theta_2 \geq 60°$ with $\theta_1 = 0°$.

4.2 Echo Disturbance

Hass[15] studied the echo disturbance when the reflected sound had the same spectrum as the direct sound, using a method of analogue simulation. The first aim of our study was the effect of changing the spectrum and the second was the effect of the direction of reflected sound approaching the listener. In an anechoic chamber, two loudspeakers were located at points in the directions $\pm 22.5°$ and 2·4 m distant from the listener and two systems—Stereo-system and Mixed-system—were used to give the test stimuli. In the Stereo-system, the listeners were informed in advance, which was the direct sound f(t), the same speech as used in the previous section, or the reflected sound $F(t) * W(t+\tau)$. In the Mixed-system, the two signals were mixed before they were fed to the loudspeakers so as to be perceived with frontal incidence by the listener. The trained twenty-three subjects were requested to judge whether the direct sound is affected by the reflected sound or not.

Three kinds of standard boundary materials were used to simulate the reflected sound, they were: a glass-fibre, a perforated board, and a rigid rectangular reflector for oblique incidence and including a perfect reflection, i.e. having no change of the spectrum. The simulated echoes were presented with level differences +10, 0, −3, −6, in dBA, and various delay times within 1–160 msec refer to the direct signal.

For example, a transfer function of a rectangular reflector calculated by Fresnel's integral at the incident angle of 20° is shown in Fig. 11. The result of subjective tests is shown in Fig. 12, compared with perfect reflection. Another example of a transfer-function of a perforated board with resonant frequency of 200 Hz at the incident angle of 60° is shown in Fig. 13, and the result of subjective tests is shown in Fig. 14. It can be seen that there is little difference between the simulated reflection from a rectangular reflector and perfect reflection, in Fig. 12, but there is quite a difference between the reflection from a perforated board and perfect reflection as shown in Fig. 14.

This fact means that we can detect the echo from a low frequency resonator much more easily than that from the other boundaries, and such echoes may be harmful to the intelligibility of speech, or the sound quality itself. We can also find rather different results for the two systems. For short delay times under 100 msec, much echo disturbance is observed in the Stereo-system compared with that in the Mixed-system. With delay times above 100 msec, we can find contrary results. It may be that, if the direction of echo is different from the direct sound, we can listen to two signals separately with long delay times.

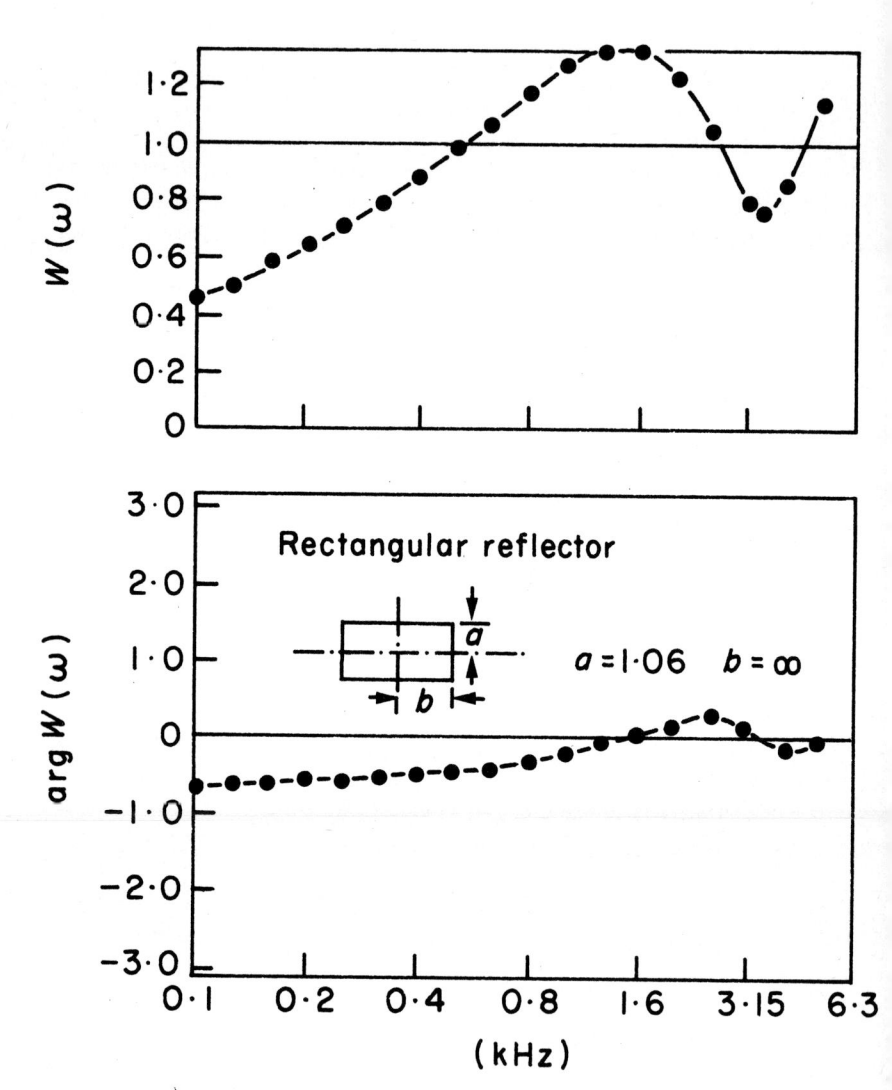

FIG. 11. Reflection transfer function of a rectangular reflector calculated by Fresnel's integral.

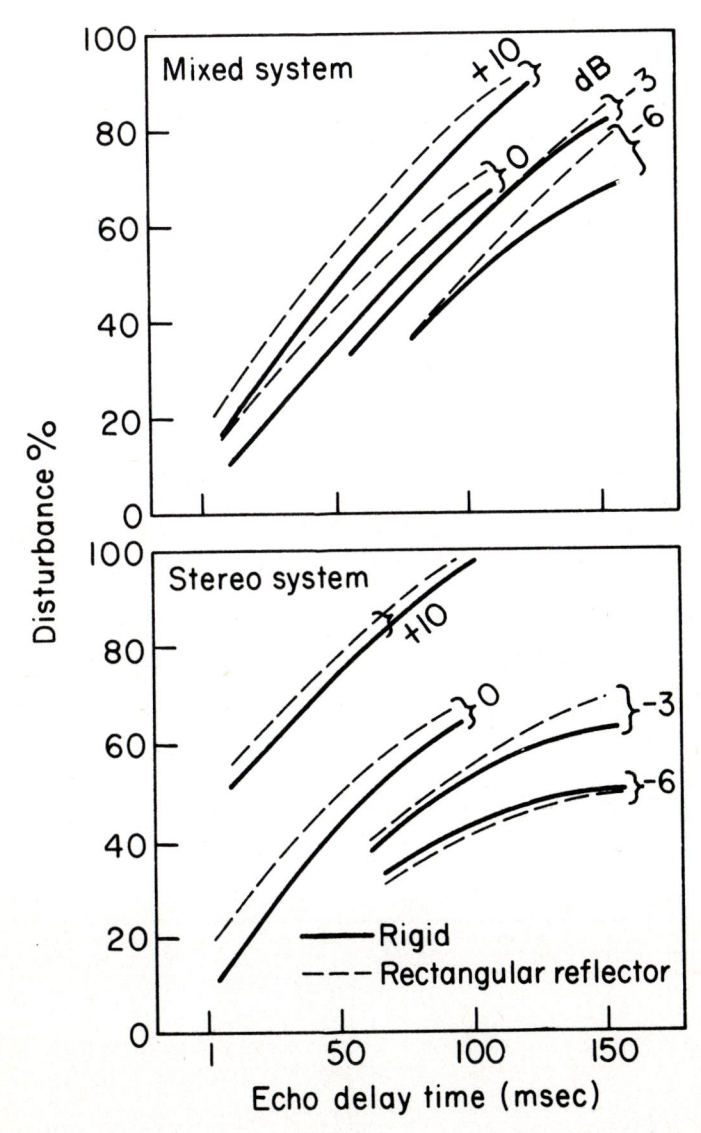

FIG. 12. Subjective disturbance by simulated echoes from a rectangular reflector as a function of delay time and at a fixed relative level of echo. —— perfect reflection (from a fixed infinite large reflector); – – – the finite sized reflector shown in Fig. 11.

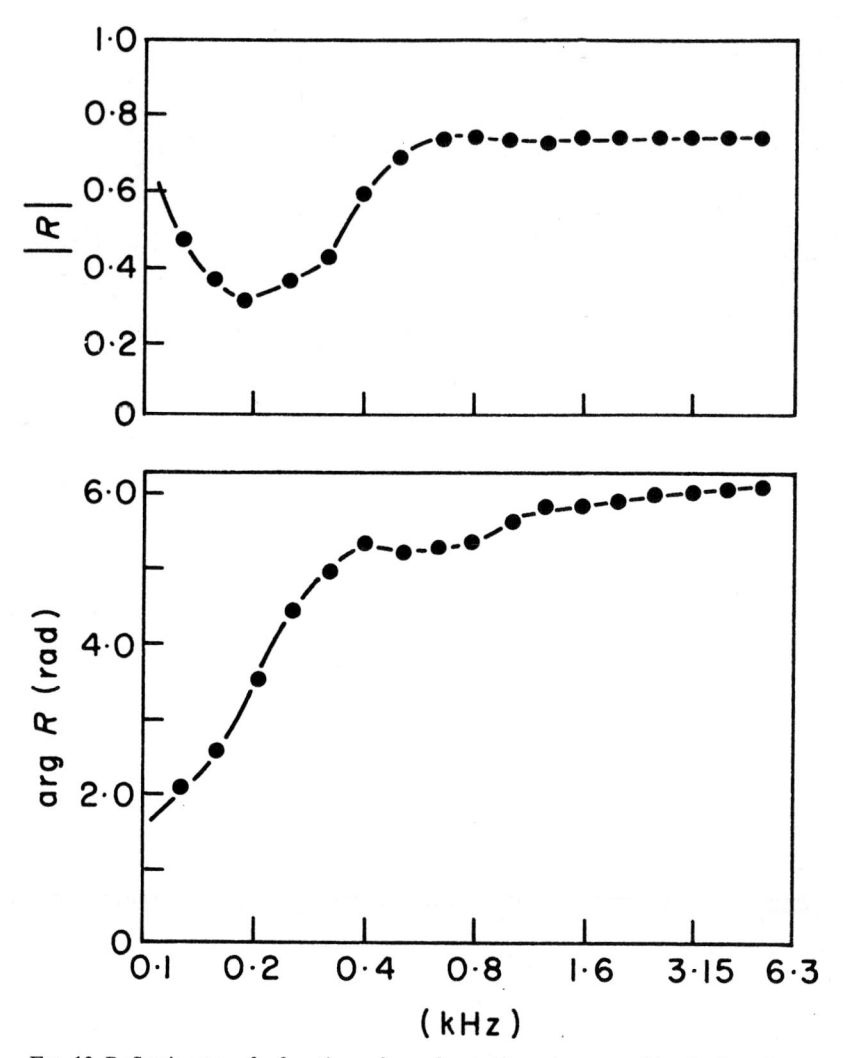

FIG. 13. Reflection transfer function, of a perforated board measured by the 'interference pattern method'.

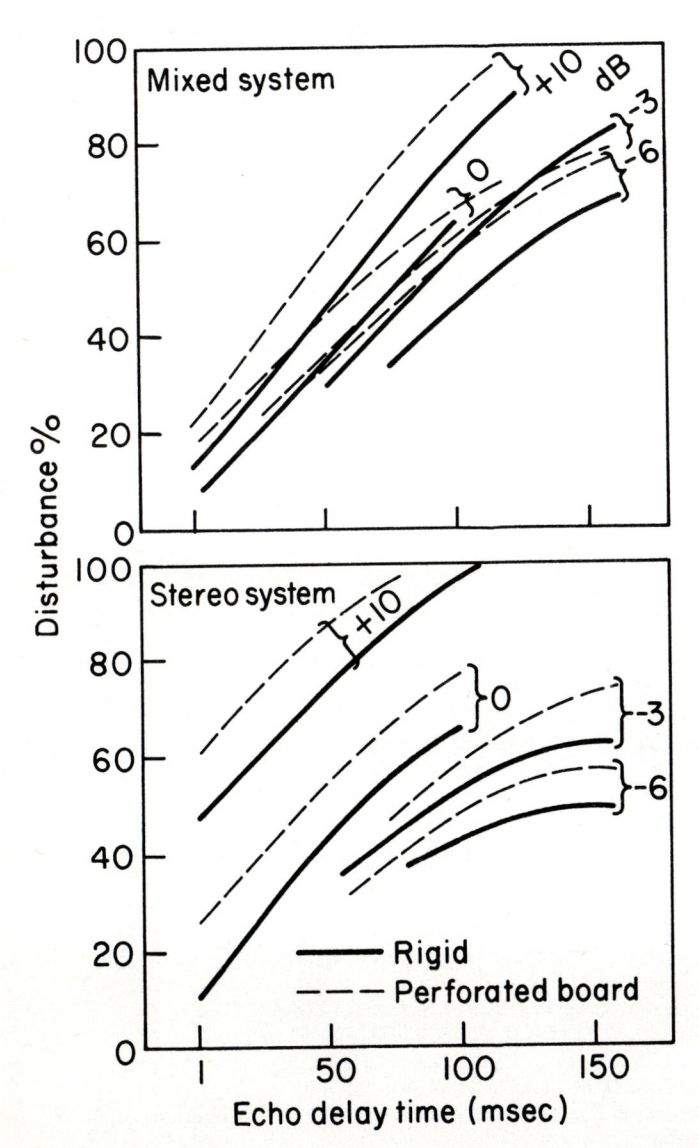

FIG. 14. Subjective disturbance by simulated echoes from a perforated board, having resonant frequency of 200 Hz as shown in Fig. 13. —— perfect reflection, same as in Fig. 12; – – – the perforated board.

5. CONCLUSION

This paper has reviewed the measurement of sound reflections from both even and uneven boundaries in terms of the transfer-functions. It has been clear that a reflection transfer-function of a boundary is necessary to simulate the sound reflection from it. And the results of two subjective tests show an important effect of the reflection transfer-function on subjective evaluations. To simulate the sound reflection by the transfer functions and using a digital computer will be a useful method to design the acoustics of a room.

The author is indebted to Mr. Y. Ando, Kobe University for preparing this report.

REFERENCES

1. For example, Knudsen, V.O. and Harris, C.M., (1950). *Acoustical Designing in Architecture*, John Wiley, p. 185.
2. Green, G., (1838). *Trans. Camb. Phil. Soc.*, **6**, 403.
3. Harris, C.M., (1945). *J.A.S.A.* **17**, 35.
4. Ingard, U., and Bolt, R.H., (1951). *J.A.S.A.*, **23**, 509.
5. Shaw, E.A.G., (1953). *J.A.S.A.*, **25**, 224.
6. Ando, Y., (1968). *Trans. Inst. Elect. Comm. Eng. Japan*, **51-A**, 303, and also *Elect. Comm. in Japan by IEEE* 51–8.
7. Ando, Y., Suzumura, Y. and Maekawa, Z., (1972). *J.A.S. Japan*, **28**, 289–98 and also (1972). *J.A.S.A.*, **51**, 155 (A).
8. Beranek, L.L.,(1962). *Music, Acoustics & Architecture*, John Wiley, New York.
9. Beranek, L.L., (1965). *Acustica*, **15**, 307–16.
10. Meyer, E., (1963). *Acustica*, **13**, 183–86.
11. Knudsen, V.O., *et al*, (1964). *J.A.S.A.*, **36**, 2328.
12. Leizer, I.G., (1966). *Soviet Physics-Acoustics*, **12**, 180–4.
13. Sakurai, Y. and Maekawa, Z., (1968). Memoirs Fac. Eng. Kobe Univ., **14**, 107–28, and *J.A.S. Japan*, **24**, 289–98.
14. Maekawa, Z. and Sakurai, Y., (1968). Proc. 6th ICA Congress, Tokyo, E 29.
15. Hass, H., (1951). *Acustica*, **1**, 49–58.
16. Muncey, R.W., *et al* (1953). *Acustica*, **3**, 168–73.
17. Schroeder, M.R., *et al* (1963). *IEEE Intern. Conv. Rec.*, **11**, 105–55.
18. Ando, Y., Shidara, S. and Maekawa, Z., (1973). *J.A.S. Japan*, **29**, 151–9. (In Japanese)
19. Schomer, P.D., (1972). *J.A.S.A.*, **51**, 1127–41.
20. Donato, R.J., (1973). *J.A.S.A.*, **53**, 589–92.
21. Escudie, B., *et al*, (1973). *Inter-Noise* **24**, 6.
22. Louden, M.M., (1971). *Acustica*, **25**, 167.
23. Maekawa, Z., (1974). 'Environmental Sound Propagation', 8th ICA Congress, London.
24. Spindel, R.C., and Schultheiss, P.M., (1972). *J.A.S.A.*, **51**, 1812.
25. de Bruijn, A., (1971). *Acustica*, **24**, 75–84.
26. Ando, Y., Shidara S. and Maekawa, Z., (1974). Proc. 8th ICA Congress, London, p. 611.

17

New Results and Ideas for Architectural Acoustics

M. R. Schroeder

*Drittes Physikalisches Institut,
Universität Göttingen,
Federal Republic of Germany.*

CONCERT HALL MEASUREMENTS

I would like to review very briefly the concert hall measurements carried out by Gottlob and Siebrasse in auditoria throughout Europe including a visit to England to measure the Royal Festival Hall. Firstly, I would like to review the method and the results very briefly and then go onto some topics, that have been suggested by this work.

It was about 1961 when B. S. Atal and I at Bell Laboratories became interested in reproducing sound fields, three dimensional sound fields, without wearing earphones in such a way that you could turn your head and the sound field would remain stationary so that you could really perceive the sound as coming from outside your head and not originating somewhere inside your head. At Erwin Meyer's Institute they had carried out experiments using many loudspeakers in an anechoic chamber. However, I thought it should be possible to do this with just two speakers. After all, we have only two ears and if one radiates two correct signals from these two speakers, then, at least for a particular location of the listener and for a particular head position of the listener relative to these speakers, we should be able to obtain the right sound pressure at the eardrums. We succeeded in doing this and the method has since been elaborated by Damaske and Mellert and used in our concert hall measurements.

Figure 1 illustrates in a very schematic way how sound reaches the two ears of a listener from two loudspeakers. For one pair of paths labelled 'S' the sound is transmitted to the same side, from the left loudspeaker to the left ear and the right loudspeaker to the right ear whilst for the paths labelled 'A' we have sound going from the left loudspeaker to the right ear and from the right loudspeaker to the left ear.

We measured these responses originally at Bell Laboratories by measuring amplitude and phase for a given configuration and different angles of incidence and then later at Göttingen. Gottlob and Siebrasse measured these responses 'A' and 'S' as impulse responses (an example is shown in Fig. 2). The upper trace, labelled s(t), is the impulse response to the 'same' side and the lower trace labelled a(t) is the impulse response to the 'other' side, and, in this case, is delayed by roughly 290 μs. The delay, of course, depends on the size of the head and the angle of incidence, in other words the direction where the loudspeaker is standing.

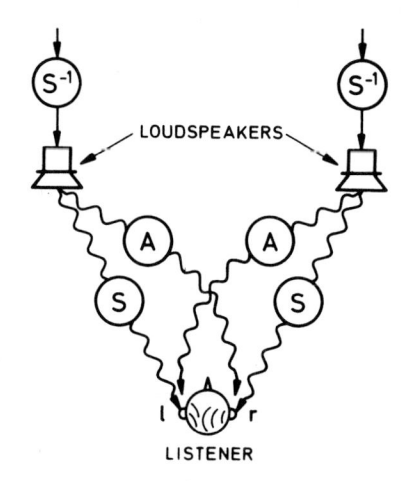

FIG. 1. Schematic diagram indicating how sound reaches the two ears of a listener from two loudspeakers.

Now in order to get rid of this a(t), this cross talk from the right loudspeaker to the left ear and the left loudspeaker to the right ear, we somehow have to do certain inverse transformations and these can only be done properly in the Fourier domain. So we Fourier transformed these responses. The result is shown in the top panel of Fig. 3 where we see the magnitude of s(t) called $|S(\omega)|$ and in the lower panel the phase as a function of frequency where a constant delay has already been subtracted because, as long as we know it, it is of no further interest. One can also see, of course, the things that Blauert and others have observed and interpreted so correctly, how the frequency response depends very much on the angle of incidence, how for a given angle of incidence you get maxima and very deep breaks which are related to the diffraction of the sound around the human head. I might also mention that for localising sound sources in the medium plane

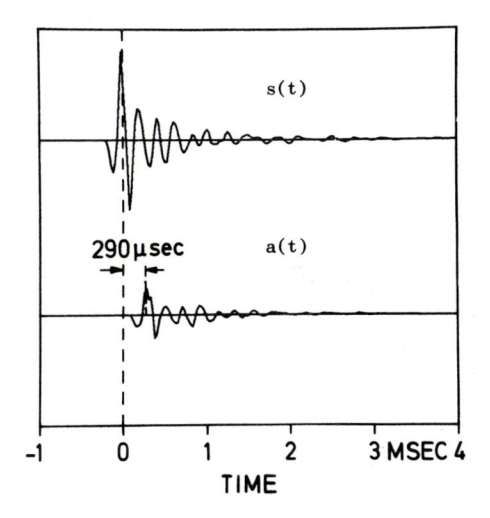

FIG. 2. Impulse responses for same side and opposite side paths as indicated in Fig. 1.

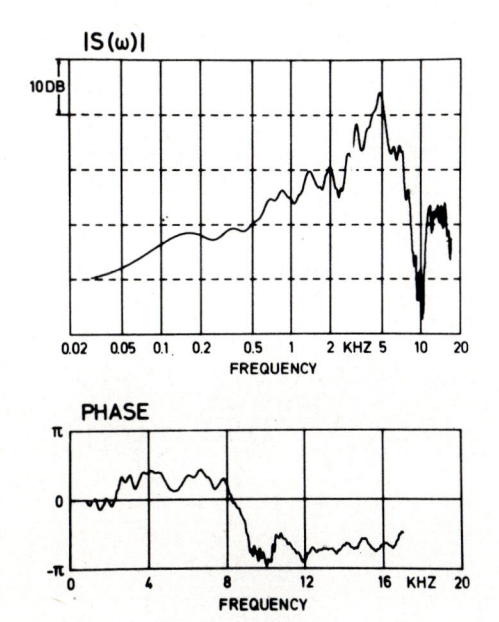

FIG. 3. Fourier transforms of the impulse indicating the magnitude (upper) and phase (lower).

where you have no delay or phase differences or intensity difference between the two ears, it is this kind of spectrum and particularly the dips in the spectrum or peaks in the spectrum that allow you to localise in the medium plane if your head is absolutely fixed (by biting into a bite-board or other method).

Figure 4 shows the kind of filter arrangement that Atal and I proposed in 1961. At the lower part of the figure you see again the two loudspeakers and these two kinds of paths, 'S' and 'A'.

Suppose you made a recording, for example in a concert hall that you want to test; you made a recording with a dummy head, an artificial head

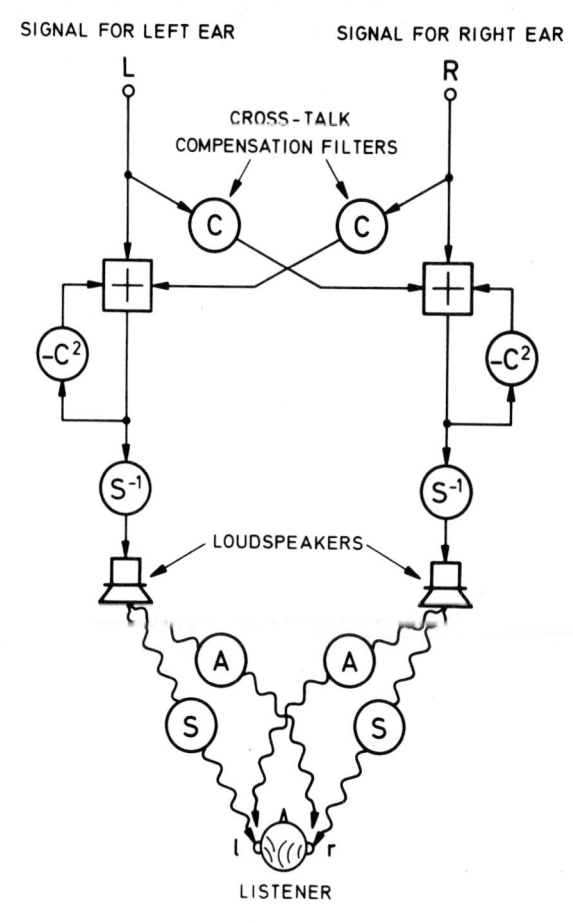

FIG. 4. Filter arrangement.

with a condenser microphone in its ears, and the signal from the left ear of course you want to get to the left ear of your listener. All right, how do you do this? You somehow have to construct, and this was our idea then, an electrical network which is the inverse of what goes on acoustically and I do not have to go into details here. 'C' is a certain filter that depends on $S(\omega)$ and $A(\omega)$ and one has to show that there is a feedback loop as you can see with a gain of C^2 and this has to be stable if you want to realise it as an analogue filter and it works nicely. In fact, it worked so well in our early demonstrations at Bell Laboratories, that when we had visitors we would ask them where they heard these sound sources coming from and when we had virtual sound sources from the side, for example $\pm 90°$, the impression was so realistic that they invariably would look in that direction to see if there was another hidden loudspeaker; of course there was none. All we used in these demonstrations was a simple delay line, again in a feedback loop to compensate for this crosstalk, and with the visitors in the right position in front of our two loudspeakers we simulated sound sources in various directions, including outside the loudspeaker base which was $\pm 22\frac{1}{2}°$. The moment they turned their heads, the illusion went away because this whole thing works only for a given head position and head directions plus or minus a few degrees.

We then obtained, many years later, in 1970, the tape from the B.B.C. with little or no reverberation on it. The B.B.C. orchestra recorded this, I understand, in an anechoic chamber and this was used by Gottlob and Siebrasse who carried out studies around Europe supported by the Deutsche Forschungsgemeinschaft. They placed their specially developed dummy head into different positions in these different concert halls and played the tape from the stage. There is a question, should one use a real orchestra or a tape? There are advantages and disadvantages for both methods. The tape has the advantage that you always have the identical material while some people claim that musicians sometimes play differently on different occasions. In addition to recording this music on the B.B.C. tape with their artificial head, Gottlob and Siebrasse also recorded impulse responses from a specially constructed 10-kV spark source. The reason for this was that we could measure objective parameters such as reverberation time, definition, something that we have, unfortunately I should say, called inter-aural coherence (because coherence really means something else but I will explain in a moment what we meant by this term). They measured these impulse responses so we could extract physical (objective) measures and so that, if at some future time we have some new ideas as to what to measure in a concert hall, we have these impulse responses and we can convolve them with music material and then do new tests with a different kind of programme material, different music or speech for example. Also, in the future, the near future I hope, we want to start changing these impulse responses to add certain modifications of these rooms that we made the recordings in—add echoes, take echoes away, change the reverberation time and so forth. This we can do on the basis of these impulse responses on the computer and then we have the modified hall represented by its impulse responses for a particular seat, for a particular collection of seats and then run the same music material through it for future tests.

Now these tape recordings were subjected to preference tests, to what we like to call 'experienced listeners' (whoever was not experienced at the beginning was experienced at the end, I am sure, because the theses took long enough, as most theses are likely to do). The subjects could always switch between two halls. They did not know which halls they were listening to and they could say, 'I prefer A' or 'I prefer B.' Now with all this information from many listeners and all possible combinations of the halls and with roughly 20 halls, many, many comparisons were collected and subjected to a factor analysis. That of course we cannot go into here but what the factor analysis does is to extract the most important factor underlying the subjective judgement. The factor analysis constructs a Euclidian space and in this Euclidian space are located the different concert halls or different seats in different concert halls. For example, if you have two seats in the same concert hall or two seats in two different halls and you look at their positions in this subjective space which the factor analysis creates, then the distance between these two seats tells you something about how different they are, and the projection of their co-ordinates in the first factor direction tells you something about which hall was more preferred. In fact in this factor analysis the first factor, the most important factor which accounts for something like 60% of the total variance, could be called the consensus preference. You see it is the factor, the underlying entity in these tests on which all subjects agreed. Factor two in such a factor analysis is the most important one on which they do not agree, where personal differences show up. Factor three is the next factor where they do not agree and so on. But factor one is the factor that they agree on, no matter how different their opinions might be in other respects and since we are interested in the acoustic quality of concert halls, we are predominantly interested in this consensus preference. The time may come in the future where we can build concert halls with different acoustics in different corners (but we are not working on it right now) and then we would need these other factors, like factor two, factor three and so forth to accommodate different musical tastes. Of course, the same goes for different musical styles which we already know from the work of Kuhl, for example, will require different reverberation times. Right now we are just trying to do something for the democratic majority, so to speak.

Now what have these tests revealed. When we looked at the concert halls with a reverberation time of less than 2·2 s we found that the most important objective parameter was reverberation time—the higher the reverberation time, the better the hall. Above 2·4 s, or something like this, again reverberation time was one of the most important parameters but with an opposite sign. The higher the reverberation time the less the hall was preferred. Now this is something, of course, that we have known for a long time, particularly again thanks to the work of Kuhl, that for the type of the music we have chosen, classical symphony (the Jupiter symphony by Mozart) a reverberation time around two seconds is optimum.

So the first thing we learned, so to speak, was something that was not really new, but we looked at a number of different parameters including the definition as defined by Meyer and Thiele in 1954, that is the energy in the first 50 ms divided by the total of the remaining energy. This they called definition and it has something to do, of course, with speech intelligibility. This too showed a very high correlation with the consensus preference, with factor one of the factor analysis. In fact, there is generally a correlation between reverberation time and definition: the longer the reverberation time, the lower the definition and vice versa. When we expressed our results in terms of definition, we did not have to distinguish between different reverberation time ranges, in other words if you want to find a subjectively meaningful measure of how the sound decays in the hall, it seems better to use definition. If you use definition to describe the decay, you get results that are the same for halls, no matter whether you have a short or long reverberation time. This is shown in Fig. 5. D1 is the first factor of the factor analysis, the consensus preference. The more a hall is on the right, if we were plotting halls here, the better it would be. But in this figure we are not plotting halls, we are just plotting the deviation from an optimum value of the definition as defined by Meyer and Thiele and you see no matter whether we look at all twenty-four halls, or nineteen halls, or eighteen halls with reverberation times less than 2·2 seconds, or twelve halls with reverberation times larger than 2 seconds, or the nineteen halls in the mid range, for all these halls we find a clear cut negative correlation which is quite significant, a high negative correlation, with factor 1, no significant correlation with factor 2. What does this mean? It means that for all the halls that we tested all subjects agreed pretty much, with a very high consistency, that the higher the definition the less the preference, for this music of course. If we had used speech, I am sure the answer would have been different.

Now I believe that we found other interesting things and before I get to these results, let me make one remark. The fact that reverberation time, or definition, turned out to be the most important factor, it also means and this is known from other applications of these factor analysis methods that

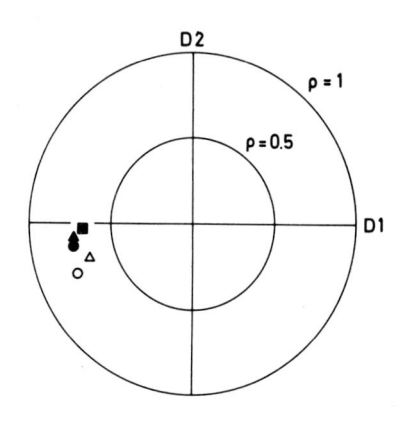

- •   24 HALLS
- ○   19 HALLS (WITHOUT THE MOST AND LEAST PREFERRED H.)
- ▲   18 HALLS : T ⩽ 2.2 SEC
- △   12 HALLS : T ⩾ 2.0 SEC
- ■   19 HALLS : 1.5 ⩽ T ⩽ 2.4 SEC

FIG. 5. Correlation: definition–preference scales.

they sort of dominate the weaker factors. It is wise to eliminate the strongest factors so that the weaker factors can play a larger role, so what we are planning to do in our future tests is to take these impulse responses and normalise the reverberation time, that is, make the reverberation time of all these 24 halls (or maybe make the definition of all these halls) equal and then hopefully the other things that we find will stand out even better. I know this has happened in other psychological tests using this kind of methodology.

Now the really new thing that we found was this inter-aural coherence which we defined as the amplitude of the maximum of the inter-aural cross-correlation function. You see, in Fig. 6 we have an impulse response from our spark source on the stage to the right ear and to the left ear so we have two time functions from which we compute on the digital computer a cross-correlation function and its highest peak. If the highest peak is very high, then we say we have high inter-aural coherence and if the highest peak in the cross correlation is low, then we say we have low inter-aural coherence. This inter-aural coherence, a truely binaural objective measure, showed the next highest correlation with subjective data and it showed a *negative* correlation as can be seen in Fig. 6. This is not quite as high as

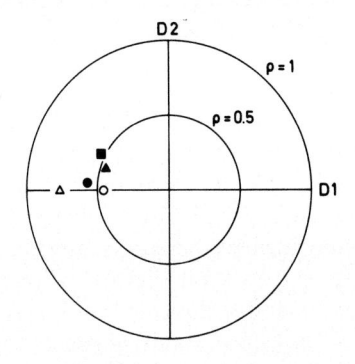

- 24 HALLS
- ○ 19 HALLS (WITHOUT THE MOST AND LEAST PREFERRED H.)
- ▲ 18 HALLS : T ≤ 2.2 SEC
- △ 12 HALLS : T ≥ 2.0 SEC
- ■ 19 HALLS : 1.5 ≤ T ≤ 2.4 SEC

FIG. 6. Correlation: inter-aural coherence–preference scales.

definition or reverberation time but you see all these halls, all 24 halls or these four different sub-groups of halls show an inter-aural coherence negatively correlated with factor one and again relatively little component or correlations with factor two, which reflects what I said before concerning inter-personal differences of our subjects. So we can conclude that inter-aural coherence as we defined it is bad for concert halls. We want concert halls in which the cross-correlation function between the two ears is small and does not have a high peak. What does it mean incidentally if the cross-correlation function between the two ears has a high peak? It means, for example, if the peak is at zero delay, that you get a lot of direct sound and little sound from other directions. Now if the peak is not at zero delay but at some other delay it means that you have a lot of sound from one direction and very little from the other direction and this seems to be bad. I do not mean I am shocked by the result, in fact, I think it is nice and it is an objective result. Now, of course, when I say that the size of the inter-aural coherence is related to whether the sound comes mostly from one direction or from many directions, I am talking about diffusion. In other words it looks as if it is good to have more diffusion in a concert hall. This may again sound old fashioned as has been said for many years but I have never seen such unambiguous objective data stating that diffusion, if inter-aural coherence is a measure of diffusion (and I think it is), is good to have in a concert hall for this type of music.

MORE DIFFUSION

Now, how do we get more diffusion in a concert hall? At this point I would like to change the subject and come to another point. Of course, we put things on the wall, all kinds of odd shapes and they will scatter the sound. But as a mathematically inclined person, I immediately ask myself, what is the optimum shape of a wall, for the case of sound arriving at normal incidence or from any other direction, so that it gets scattered as widely as possible? For example, if I hit the wall, normally I would like, in the ideal case, to have the sound evenly scattered in all directions both in the hori-

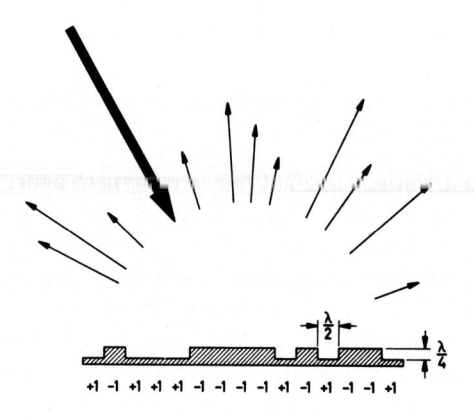

FIG. 7. Pseudo-random maximum-length code $N = 15$.

zontal plane and the vertical plane. Is this possible? The answer is, it is possible at one frequency. I will indicate later how we can do this now for many octaves and of course we are interested in a wide frequency range. But first, how do you do this for a single frequency?

In computer science, there is something called maximum-length codes; these are used to create pseudo-random noise on computers. These are certain sequences of ± 1. The magnitude of their Fourier transform—their power spectrum—is as flat as possible, in fact it is exactly flat. It is a noise which has an extremely flat spectrum except for the DC component (as opposed to Gaussian noise which itself has a sort of Gaussian spectrum for any finite time interval). In Fig. 7 you see such a pseudo-random maximum-length code for $N = 15$. Their length N is always one less than a power of two but they can be arbitrary long and you see the code is $+1$, -1, $+1$, $+1$, $+1$ and so forth. Now there is a relation between the power spectrum of spatial functions and their radiation pattern or, if it is a spatial function of reflection coefficients along the x-axis, then the reflection pattern is related to the Fourier transform—to the power spectrum of this sequence. A wide flat power spectrum means that you get a wide distribution of reflection angles. I do not think I have to go here into the details of this relationship between angles and power spectra of spatial functions, it is a simple fact. In other words, if we could shape a wall to have reflection coefficients, $+1$, -1, $+1$, $+1$ and so forth for a given frequency then even if we hit this wall from a single direction we would expect to have fifteen or fourteen angles of reflection all with equal intensities and this is illustrated here.

Now how do we get reflection coefficients of ± 1? Well, we just make notches in the wall that are a quarter wavelength deep. At the wall the sound is reflected with a reflection coefficient of $+1$. At a notch a quarter wavelength deep it is reflected with a reflection coefficient of -1, so that is how you obtain the reflection coefficient of -1. So if we shape a wall like this we should get an optimum scattering of sound. Ideally, these properties of the maximum-length codes are true only if they are periodically repeated an infinite number of times but they should work fairly well also for a single code.

I asked Mr. Henze of my staff at Göttingen to shape a piece of sheet metal according to this code with quarter wavelengths steps and irradiate the sheet with three-centimetre microwaves, and measure the radiation pattern which should be a very broad one. Well the result was anything but broad, in fact it is a very sharp reflection. However, I knew the mathematics was correct so I asked the person who made the panel, how deep did you make these steps, and he said $\lambda/2$, but of course when you make them $\lambda/2$ then you get a reflection coefficient of 1 everywhere. So I asked for a new piece of sheet metal with $\lambda/4$ to be made. In Fig. 8 you can see the piece of metal with $\lambda/2$, the wrong part, on the bottom whilst the top part is the $\lambda/4$ design.

Figure 9 is the reflection diagram for a 3-cm wavelength with an incident wave from above. How sensitive this is, though, to the correct code is illustrated by Fig. 10 where we have almost the same maximum-length code. This was the idea of Mr. Henze who made the panel. He put a little metal

FIG. 8. The metal sheets used in the experiments. The upper sheet with $\lambda/4$ steps and the lower sheet with $\lambda/2$ steps.

FIG. 9. The reflection pattern from $\lambda/4$ stepped surface.

strip above one of the grooves (he did this behind my back; he always likes to check whether these theories mean anything) and lo and behold this little change already gives you a reflection pattern, Fig. 11, that is much more concentrated.

The question is can these principles be applied to a wide range of frequencies and the answer is a qualified yes. A student at our Institute at Göttingen (Mr. von Heessen) is working on models for acoustic waves. Incidentally, in order to make this work for a wide frequency range, one should exploit the second dimension. So far we have scattered only in the horizontal, in one dimension. Now we are going to two-dimensional rays based on similar codes to scatter into the whole solid angle and I believe that we will be able to get a very good scattering over four or five octaves which I think is probably all we need.

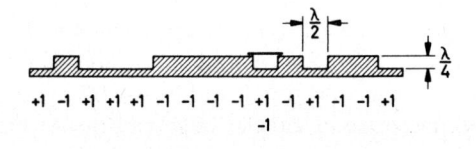

MAXIMUM LENGTH CODE
N = 15

FIG. 10. Section through metal sheet with strip applied to one of the grooves.

FIG. 11. Reflection pattern from surface with metal strip attached.

MEASURING CONCERT HALLS WITH MUSIC

To conclude with another topic and something that we have learned from these measurements. The tests that I reported on in the first part of my talk were carried out in empty halls and of course many modern halls today have similar acoustics, whether they are occupied or not, because the seats are designed that way. However, some of the older halls like the Vienna Musikvereinssaal are not so designed and, whatever you measure in these halls, you cannot really say anything about the acoustical qualities since they are not measured with an audience and I did not feel it was justified to make these measurements with audiences because we were still trying out the method. How could I ask city architects or educational officials to drum up three thousand school children for methods and measurements that perhaps were not worth anything?

Now we have more confidence and we can afford to measure in halls with an audience. On the other hand I think this problem will always be with us, how can we make significant measurements with a full audience without firing pistols and cannons or white noise and I have asked myself, is it possible to measure a concert hall with *music*? We just go to a concert and we put a few microphones on the stage and record the orchestra on the stage and we have one or several dummy heads sitting in the audiences and record from them. Is there a possibility to take the signals from the stage and take the signals from the dummy heads and compute the response from the stage to the audience even if it takes a year of computing? I think the answer to this question is yes and that is illustrated below.

Concert Hall measurement with music

	Measured Signals	Fourier Transforms
Stage	$s(t)$	$S(\omega)$
Audience	$r(t) = a(t) + n(t)$	$R(\omega) = A(\omega) + N(\omega)$
'Wanted' impulse response	$h(t)$	$H(\omega)$

$$H(\omega) = \frac{A(\omega)}{S(\omega)} = \frac{R(\omega) - N(\omega)}{S(\omega)}$$

Note that the audience signal which I call $r(t)$ consists of two parts $a(t)$ which is a signal that arises from the stage plus a noise signal $n(t)$ which is noise arising from air-conditioning equipment, from the audience moving or even changing temperature gradients. Anyhow, there are unwanted effects which I collected in this n(*t*) and its Fourier transform N(ω).

Now what we want is the impulse response from the stage to the two ears of our dummy head or its Fourier transform H(ω)—it makes no difference, we can always go back and forth between these two by Fast Fourier transform. H(ω) is simply given by A(ω) divided by S(ω). But we cannot measure A(ω), we can only measure R(ω) and we do not know what N(ω) is. Well, there are two problems here. We cannot simply Fourier transform the audience signal and then divide this by the Fourier transform of the stage signal and expect it to be the impulse response. We have to take noise into account. The second problem is that we cannot really divide by the Fourier transform of the stage signal, S(ω), because it will have zeroes or very low values and one thing that is forbidden is to divide by zero.

So how do we solve these problems? One can do a maximum likelihood approach, in other words I can measure the received signal at the dummy head and on the stage and H(ω), the Fourier transform of the impulse response, can be written as follows:

$$\text{maximum likelihood solution} \quad H(\omega) = \frac{\Sigma R(\omega) S^*(\omega)}{\Sigma |S(\omega)|^2}$$

Σ : averaging over small frequency ranges (*ca.* 1 Hz)
OR : averaging over different time segments (*ca.* 10 s)
$R(\omega)S^*(\omega)$ = Fourier transform of cross-correlation between r(*t*) and s(*t*)
$|S(\omega)|^2$ = Fourier transform of auto-correlation of s(*t*)

The sum over $|S(\omega)|^2$ is the saving grace here; it avoids dividing by zero. I would record for example one larger musical piece lasting, say, a thousand seconds. This means, in the Fourier transform, I get one point every 1/1000 Hz, but I do not need the information for every 1/1000 Hz, I can now average over one thousand frequency points because I only need the information every one Hertz so the summation could be over such frequency points. Alternatively, the averaging could be over different time segments. In other words, instead of recording a whole musical piece and Fourier transforming it, I can record it in one piece but then chop it up into segments of a few seconds length and then transform them individually and then average. The trick here is that in the denominator we have a sum with many terms over a positive quantity and now the probability that I will be dividing by a small number or by zero, can be made arbitrary small.

The only problem that remains now is computation. If I record one thousand seconds of music and sample 40 000 times a second and Fourier transform it, that is 40 000 000 points or two hours of computation for just one seat. Figure 12 shows how we think we can do this more efficiently together with a small computer that does most of the computations in real time. We have a microphone on the stage and we have an all-pass filter to make whatever we record on the stage more Gaussian. We scramble it because once we have a Gaussian-like signal, we can clip it and then put it

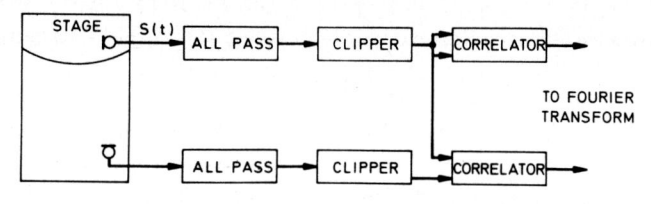

FIG. 12. Practical implementation with *one*-bit correlators.

into a one-bit correlator which can work in real time since we have one correlator for the auto-correlation and one correlator for the cross-correlation between stage and audience signals. So most of the required calculation is done by a fast, small, digital computer at one-bit accuracy. I think it can be done, although this remains to be seen. Then the remaining Fourier transform will have only a thousand points or so and then regular computers can take over and plot out impulse responses.

18

The Correlation between Subjective and Objective Data of Concert Halls

H. WILKENS and G. PLENGE

Heinrich Hertz Institut,
Berlin-Charlottenburg,
Federal Republic of Germany.

It is the aim of any investigation in architectural acoustics to contribute to the design of concert and similar halls so as to ascertain a satisfactory hearing impression. Correlation between physical data describing a sound field in a hall and the hearing impression perceived there can only result from investigation of existing halls.

A first attempt to describe hearing impression was made by Beranek[1] and also Parkin et al[2], but they asked subjects for their memories of many halls in which they had been listeners. There was different music being played in those halls and there were long intervals between listening in one hall and another. Kuhl[3] and Somerville[4] therefore made investigations using recordings instead of in the hall so that the subjects who judged the recordings did not know anything about the location of the recording. Now we know that a re-creation of the original hearing impression is attainable only by head-related stereophony[5-11] as done by Gottlob and Siebrasse[20] and also in Berlin[7].

In the investigation of existing halls, it has to be decided what kind of sound source and signal should be applied to excite a room. In some of the investigations hitherto made one or several loudspeakers were used as exciting sound sources, and the signals used were musical passages recorded with a low degree of reverberation. One of the difficulties of the investigation lies in the reproduction loudness. While it can be assumed that a concert hall as transmission system is independent of loudness, this is certainly not the case with the human ear. It is, on the contrary, very probable that the loudness of signals has some influence on the hearing impression.

Furthermore it must be questioned whether an extended sound source composed of various individual sound sources can be replaced by a few loudspeakers. To answer the question whether loudspeaker excitation produces a different evaluation of a location for listening from excitation by an orchestra, two recordings of a passage from Mozart's Jupiter Symphony were made at four different seats in the Berliner Philharmonie. The first one was a live recording during an orchestra rehearsal and the second one was taken from a loudspeaker reproduction[12]. Both recordings were made with artificial heads at the same seat.

To examine the influence of reproduction loudness in loudspeaker excitation the latter recording was reproduced with two different degrees of reproduction loudness. As the dynamic range of the record is very limited, both its forte and piano passages were made equal to the apparent loudness of the live performance. All the recordings were then presented to subjects for an evaluation using a questionnaire, as shown in Fig. 1.

We only show the results of three bipolar rating scales Figs. 2–4. In the case of the bipolar rating scale (large–small) the mean value of the answers of 20 persons tend to the 'large' end in the case of orchestra excitation and

		1	2	3	4	5	6	
1	small	---	--	-	-	--	---	large
2	pleasant	---	--	-	-	--	---	unpleasant
3	unclear	---	--	-	-	--	---	clear
4	soft	---	--	-	-	--	---	hard
5	brilliant	---	--	-	-	--	---	dull
6	rounded	---	--	-	-	--	---	pointed
7	vigorous	---	--	-	-	--	---	muted
8	appealing	---	--	-	-	--	---	unappealing
9	blunt	---	--	-	-	--	---	sharp
10	diffuse	---	--	-	-	--	---	concentrated
11	overbearing	---	--	-	-	--	---	reticent
12	light	---	--	-	-	--	---	dark
13	muddy	---	--	-	-	--	---	clear
14	dry	---	--	-	-	--	---	reverberant
15	weak	---	--	-	-	--	---	strong
16	emphasised treble	---	--	-	-	--	---	treble not emphasised
17	emphasised bass	---	--	-		--	---	bass not emphasised
18	beautiful	---	--	-	-	--	---	ugly
19	soft	---	--	-	-	--	---	loud
	Name:		Date:					Item no.:

FIG. 1. Semantic differential for evaluation of different situations in concert halls.

to the 'small' end in the cases of loudspeaker excitation. Furthermore there are different judgements for different seats in the latter case. We obtained similar results for the bipolar rating scale (dry–reverberant). For the bipolar rating scale (clear–unclear) there are similar results for both kinds of excitation but different judgements for different seats in the case of orchestra excitation and no differences for loudspeaker excitation.

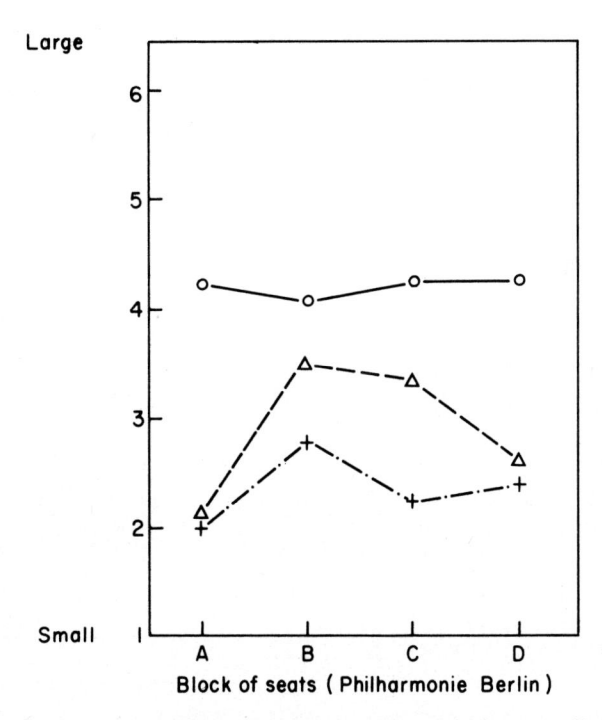

FIG. 2. Mean values of evaluations on the bipolar rating scale (large–small) for different excitations of the concert hall.
  ○ — ○   orchestra excitation;
  △ — — △   loudspeaker adjusted to equal orchestra forte;
  + — · — +   loudspeaker adjusted to equal orchestra piano.

If the more difficult use of an orchestra is chosen as an (exciting) sound source for the acoustical evaluation of rooms, the question arises whether an orchestra can produce music in different rooms of such a high degree of congruence that acoustical differences between the rooms can be made out with certainty by the hearer. To settle this question various recordings of music were recorded at one and the same seat and one record was recorded at various seats. This proved that differences in the recording position can be expected to be perceived with certainty, but not so differences in interpretation[13]. In our view, the results supply an affirmation of the technique of recording one orchestra playing in various rooms. This was

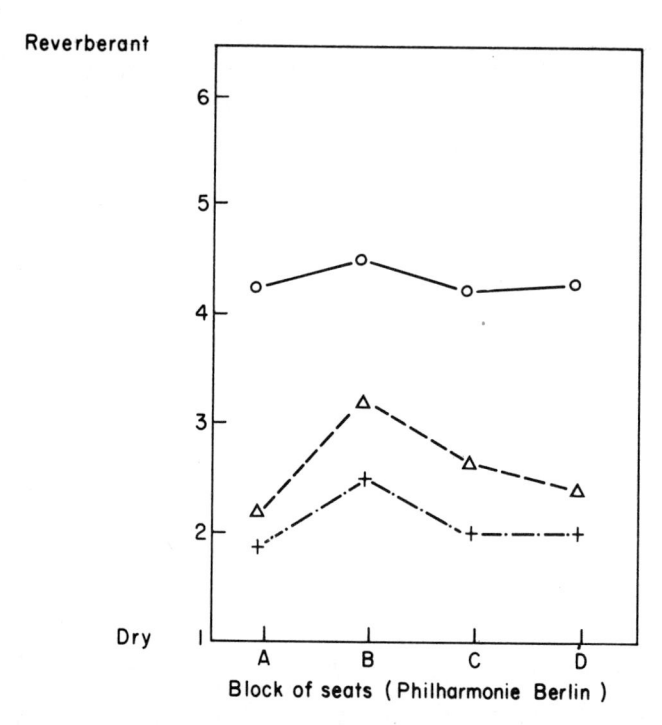

FIG. 3. Mean values of evaluations on the bipolar rating scale (dry-reverberant) for different excitations of the concert hall.
○ — ○   orchestra excitation;
△ — — △   loudspeaker adjusted to equal orchestra forte;
+ — · — +   loudspeaker adjusted to equal orchestra piano.

made possible by a tour which the Berlin Philharmonic Orchestra conducted by H. v. Karajan made to several West German cities[14]. Although the concert halls were determined by the tour schedule, a considerable variety of rooms was tested. Investigations were carried out in the halls shown in Fig. 5.

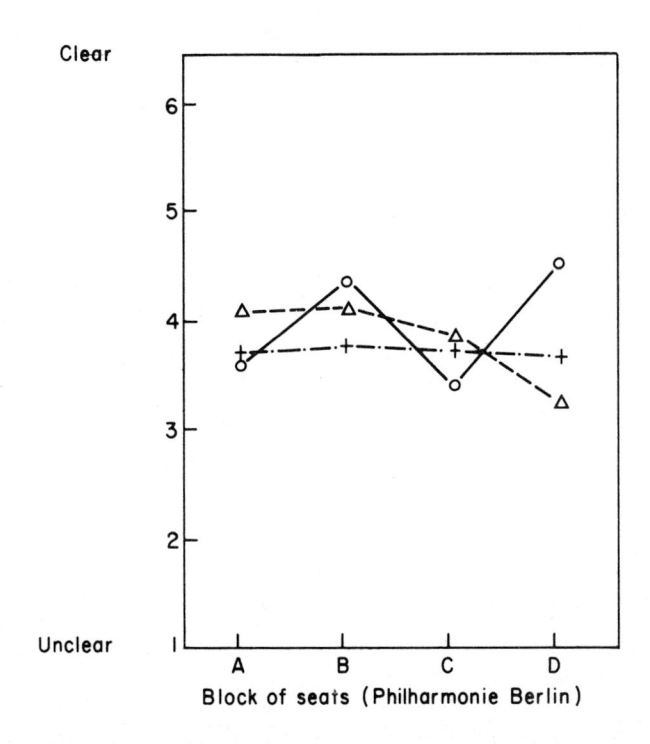

Subjective and Objective Data of Concert Halls 217

FIG. 4. Mean values of evaluations on the bipolar rating scale (clear–unclear) for different excitations of the concert hall.

○ — ○ orchestra excitation;
△ — — △ loudspeaker adjusted to equal orchestra forte;
+ — · — + loudspeaker adjusted to equal orchestra piano.

Hall	Form	Seats	Volume	$T_{500-1000 \text{ Hz}}$
Berliner Philharmonie	Arena	2200	$\approx 25\,000 \text{ m}^3$	2·0 s
Musikhalle Hamburg	Rechteck	1980	$\approx 11\,600 \text{ m}^3$	2·2 s
Stadthalle Hannover	Rundbau	3660	$\approx 34\,000 \text{ m}^3$	2·0 s
Stadthalle Braunschweig	Sechseck	2166	$\approx 19\,000 \text{ m}^3$	1·9 s
Rheinhalle Düsseldorf	Rundbau	1842	$\approx 33\,000 \text{ m}^3$	2·5 s
Stadthalle Wuppertal	Rechteck	1614	$\approx 25\,000 \text{ m}^3$	2·7 s

FIG. 5.

In all these halls seats had to be chosen for the investigation with corresponding aspects. Following are the characteristics of the places that were chosen.
1. One seat was to be in the first third of the hall, but not too close to the orchestra.
2. One seat was out in the middle of a hall rather close to a side wall.
3. One seat was chosen in the back third of each hall at an adequate distance from the rear wall.
4. One seat was located on the balcony, a little away from a central position.

The allocation of the recording seats in a rectangular hall is shown in Fig. 6. The allocations in the other halls were made with corresponding placings.

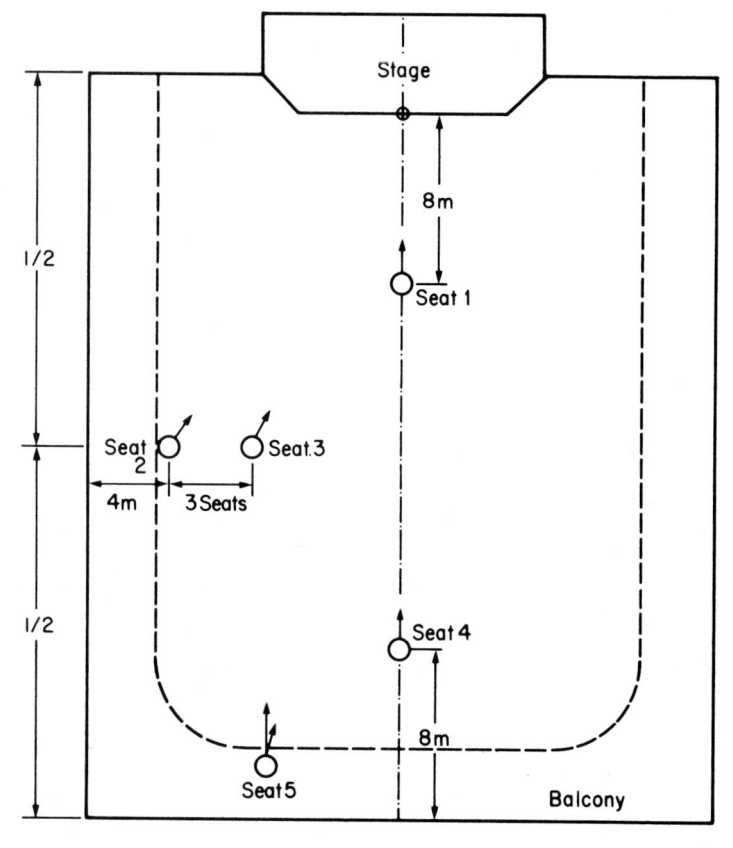

FIG. 6. Positions of artificial heads in a concert hall (fictive rectangular hall).

To determine the dependence of an evaluation of a room on the kind of music[15], a work each from the Classical, the Romantic and the Recent periods were chosen, each of them played by a full orchestra. The passages were:

Mozart, W. A.   Jupiter Symphony—1. Movement
Brahms, J.   1. Symphony—4. Movement
Bartok, B.   Concerto for Orchestra—1. Movement.

The selections were made so as to present tutti, wind and string passages. Further, the groups of instruments were to play alternatively and contrapuntally with various degrees of loudness and in various tempi. The selections had to be long enough to enable the listeners to adapt to the acoustical properties of the specific hall, which could be expected to take longer than in investigations of the perceptibility of difference.

In an initial investigation[14] the recorded passages were presented to the subjects in pairs for comparison with respect to preference. This brought out differences of judgement which could be interpreted as taste differences.

To shed light on the complex process of evaluation, the data obtained in the investigation must be useful in disclosing various aspects of evaluation. One method of formulating such data is known as 'semantic differential' or 'set of bipolar rating scales'[16], as was done by Hawkes & Douglas[17] and Sproson & Burd[18]. According to this method, each individual recording is assessed by means of several verbal bipolar rating scales. As the assessment is carried out by choosing from a number of given expressions, it is highly important to find adequate expressions. These were developed with the help of a number of musicians, sound recording engineers and acousticians. They were requested to make lists of expressions that they considered important for the description of hearing impressions. A questionnaire was developed from the multitude of expressions thus obtained and tested as to applicability[19] (see Fig. 1).

Subsequently 40 subjects evaluated 60 recordings using the improved questionnaire. As an average evaluation was aimed at, the individual subjects' evaluations were averaged and intercorrelated over the bipolar rating scales. The correlation matrix was then used as the basic material of a factor analysis. The degree of variance accounted for by the factors is shown by the following Fig. 7. It appears that 89% of the existing judgement variance can be accounted for by three factors, and that no additional factors are likely to make any considerable contribution to the analysis. This analysis, however, does not include quality judgements, since the distribution of quality judgements for some of the recordings has two maxima. However, as the quality of the individual recordings and the question of how it can be integrated into the established set of factors is well worth studying, the subjects' quality judgements were examined individually.

One of the results of the pair comparisons mentioned above is that differences in taste come into play. Summing up the results one could say that there are people who prefer a full sound and others who prefer a light, transparent sound. As an evaluation of this, we examined the correlation between quality judgements and the subjects' decisions regarding the (large–small) bipolar rating scale. This resulted in a clear division of subjects into two groups. Now, if the quality judgements of both these groups are averaged dividing the range of judgements as to the (large–small) scale into three sections, the dependence shown in Fig. 8 appears.

Of course, if two factor analyses are carried out, one for each group, of all the bipolar rating scales including the quality judgements, they result

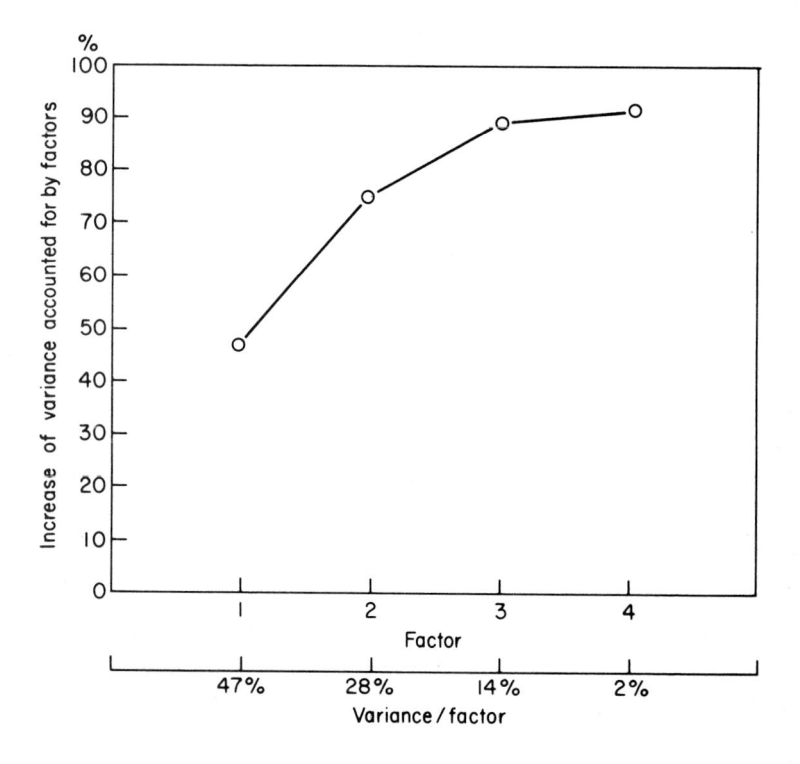

FIG. 7. Dependence of increase of variance accounted for by factors against number of factors. Experiment with 40 persons and 60 tests.

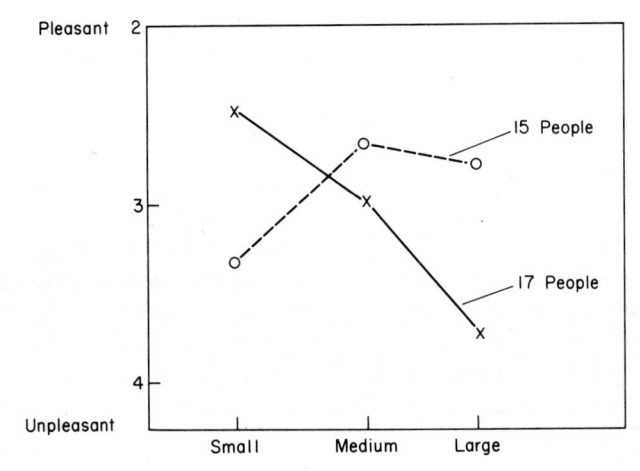

FIG. 8. Mean evaluations of the two groups on the bipolar rating scale (pleasant–unpleasant) against standardised evaluations on the bipolar rating scale (large–small) for all test examples.

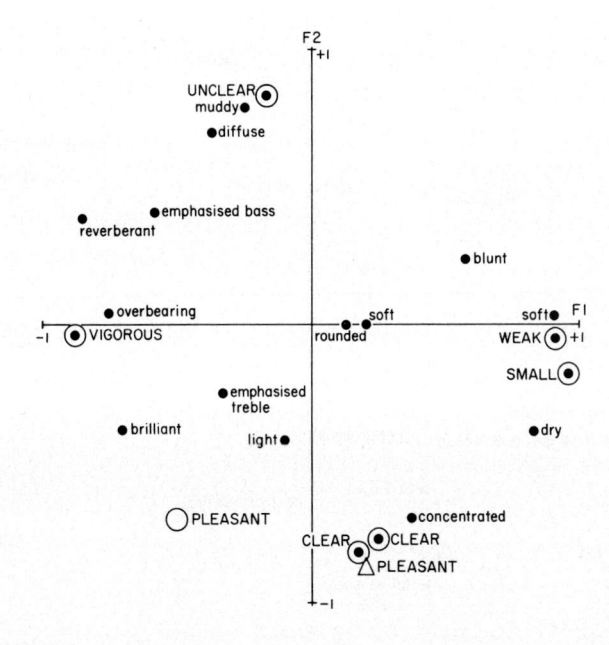

FIG. 9. Representation of Factor Plane F1–F2. The bipolar rating scales taken for the interpretation of factors indicated ●. The quality evaluations of the two groups are indicated △, ○.

in two slightly different factor spaces. Again only three factors can reasonably be extracted, that is to say, quality judgements do not represent independent aspects of evaluation.

To make out equalities or at least similarities between the two factor spaces of the two groups of subjects, a comparison of factor structures is required. The result of such a comparison of factor structures can be demonstrated by similarity. It appears that the two factor spaces are almost equal (similarity coefficient *ca.* 0·96) except for the quality judgements. This means that the assessment of hearing impressions is made equally by all the subjects, except that they apply different criteria for the assessment of quality.

The next task is to integrate the quality judgements of the two groups into the factor-space which has been found valid for both groups. This overall factor space is shown by Figs. 9 and 10 where:

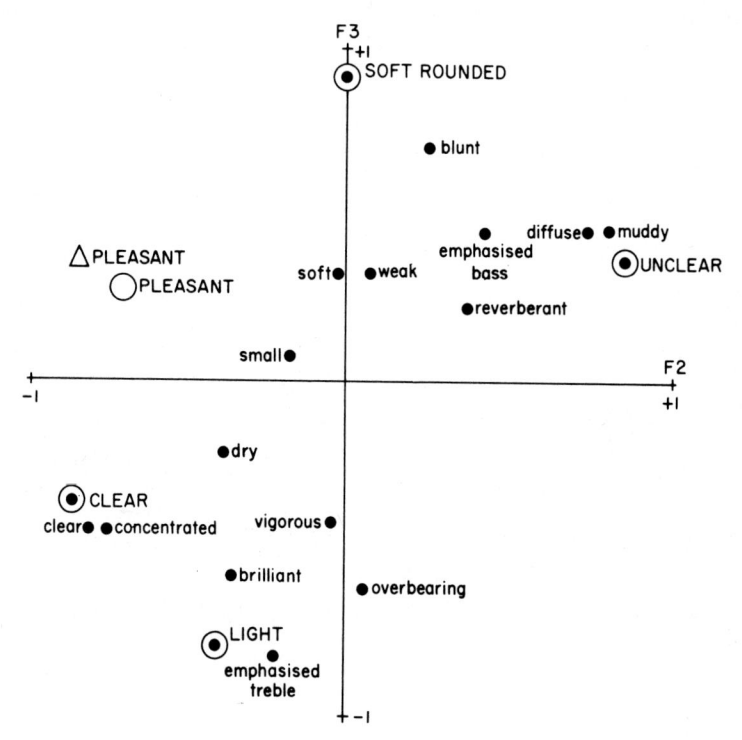

Fig. 10. Representation of Factor Place F2–F3. The bipolar rating scales taken for the interpretation of factors are indicated ●. The quality evaluations of the two groups are indicated △, ○.

F 1 = sensation of sound force
F 2 = sensation of definition
F 3 = appreciation of sound.

It can be stated that the sound fields shown are evaluated roughly along the same lines by all the subjects, but they will presumably come to different quality judgements based on differently applied criteria.

If we compare these results with the results of Gottlob and Siebrasse[20] we can perhaps say that their first factor is comparable with our second factor. Also in our investigations people agree on this aspect.

Another highly important conclusion can be drawn from the determination of factor spaces, namely that loudness perceived at the place where it is heard, which is certainly one component of the sensation of sound force, has an essential influence on quality judgements.

The other component could be the source spread or the volume.

Since it has been found in this analysis that any evaluation of hearing impressions is made essentially using three criteria independent of each other, it appears desirable to characterise all the evaluated recordings by data concerning the three aspects. These data will supply a reliable description of the subjective hearing impressions. These data can be gathered from the estimation of factor scores.

Another part of such investigations is to describe these impressions by physical data. This is an outline of how we proceeded to achieve this: impulse measurements were carried out on the recording seats[21]. The data thus obtained were evaluated in accordance with several principles of evaluation and subjected to a factor analysis for reduction of the data. It appeared that the initial reverberation data were highly correlated with the reverberation time. Another linearly independent set of data described proportions between amounts of energy applied initially and later on. After the factor analysis of the physical data the factor scores of all the recording seats for three different frequencies (500 Hz, 1 kHz, 2 kHz) were estimated. Thus each recording seat can be described by six data calculated by the impulse measurements.

Now we have to look for connections between physical and subjective data sets describing the properties of the recording seats in the different halls. To correlate both, the subjective and objective sets of data, the procedure of canonical correlation was applied[22]. This procedure comprises besides the factor analyses, to which both sets of data were submitted, a rotation of both the sets resulting in a maximum of covariance between values of the respective axes of the multi-dimensional spaces. Thus connections between both sets of data can be revealed.

A first analysis did not show any connections which were not of lesser significance. This, however, is not surprising with the sensation of sound force being one of the aspects of subjective evaluation. Corresponding data are not contained in the criteria, as they were independent of loudness. The assessment of connections existing between subjective and objective data can be made more adequate by including into the evaluation a measure of loudness obtained by an analysis of frequency level distribution. This allows for an explanation of 50% of the variance in evaluations.

It appears that the objective criteria hitherto applied do not allow for a satisfactory description of hearing impression. Therefore we must try to improve the physical data to describe the hearing impression represented by the subjective data.

We can summarise: the presented sound fields are evaluated roughly along the same lines by all the subjects, but they will presumably come to different quality judgements based on differently applied criteria.

These three criteria are: F1 sensation of sound force; F2 sensation of definition and clearness; F3 appreciation of sound.

Another highly important conclusion can be drawn from the determination of factor spaces, namely that perceived loudness has an essential influence on quality judgements at the place where it is heard.

Another result is that we do not have to look for the *one* optimal hall with the optimal RT, the optimal diffusity and so on, but—as Cremer[23] suggested 20 years ago—to point out the boundaries which limit the optimal range within which physical data can fall. This, on the other hand, leads to one great advantage—many halls with different sizes and shapes can accomplish these requirements.

REFERENCES

1. Beranek, L., (1962). *Music, Acoustics and Architecture*, John Wiley, New York and London.
2. Parkin, P.H., Scholes, W.E. and Derbyshire, A.G., (1952). 'The Reverberation Time of ten British Concert Halls', *Acustica*, **2,** 97.
3. Kuhl, W., (1954). 'Über Versuche zur Ermittlung der günstigsten Nachhallzeit grosser Musikstudios', *Acustica*, **4,** 618.
4. Somerville, T., (1953). 'Subjective Comparisons of Concert Halls,' *BBC Quarterly*, **8,** 125.
5. Damaske, P. and Wagener, B., (1969). 'Richtungshörversuche über einen nachgebildeten Kopf,' *Acustica*, **21,** 30–5.
6. Kürer, R., Plenge, G. and Wilkens, H., (1969). 'Correct spatial sound perception rendered by a special 2-channel method,' 37th AES Convention, New York, No. 666 (H-3).

7. Wilkens, H., (1971). 'Beurteilung von Raumeindrücken verschiedener Hörerplätze mittels kopfbezogener Stereophonie,' Proceedings of 7th ICA (Budapest 1971), Vol. 3, p. 709.
8. Wilkens, H., (1972). 'Kopfbezügliche Stereophonie—ein Hilfsmittel für Vergleich verschiedener Raumeindrücke,' *Acustica*, **26**, 213.
9. Plenge, G. and Romahn, G., (1972). 'The Electro-Acoustic Reproduction of "Perceived Reverberation" for Comparisons in Architectural Acoustic investigation,' *JASA*, **51**, 421.
10. Wettschureck, R., (1973). 'Die absoluten Unterschiedsschwellen der Richtungswahrnehmung in der Medianebene beim natürlichen Hören, sowie beim Hören über ein Kunstkopf-Übertragungssystem,' *Acustica*, **28**, 197.
11. Wettschureck, R., Plenge, G. and Lehringer, F., (1973). 'Entfernungswahrnehmung beim natürlichen Hören sowie bei Kopfbezogener Stereophonie,' *Acustica*, **29**, 260.
12. Burd, A.H., (1969). 'Nachhallfreie Musik für akustische Modelluntersuchungen,' *Rundfunktechnische Mitteilungen*, **13**, No. 5, 200.
13. Plenge, G., (1972). 'Über einige Probleme bei der elektronischen Vermittlung von Höreindrücken aus Räumen,' DAGA-Tagung (Stuttgart 1972), Berichtsheft VDI, Düsseldorf p. 154.
14. Wilkens, G., (1972). 'Vergleich der Hörsamkeit verschiedener Konzertsäle mittels Kopfbezogener Stereophonie,' DAGA-Tagung (Stuttgart 1972), Berichtsheft VDI, Düsseldorf, p. 158.
15. Reichardt, W., (1968). 'Der Impulsschalltest und seine akustische Bedeutung,' 6th ICA Congress, Tokyo 1968, GP-2-2.
16. Osgood, L.E., Suci, G.J. and Tannenbaum, P.H., (1967). *The measurement of meaning*, Illinois Press, Urbana.
17. Hawkes, R.J. Douglas, H., (1971). 'Subjective Acoustic Experience in Concert Auditoria,' *Acustica*, **24**, 235.
18. Sproson, W.N. and Burd, A.H., (1974). 'Analysis of Factors Relating to Acoustic Quality derived from Recordings made in a model Studio,' 8th ICA, Vol. II, London, p. 614.
19. Wilkens, H., (1973). 'Über die Brauchbarkeit eines Polaritätsprofils zur Beschreibung verschiedener Raumeindrücke,' DAGA-Tagung (Aachen 1973), Berichtsheft VDI, Düsseldorf, p. 415.
20. Gottlob, D. and Siebrasse, K.F., (1973). 'Vergleichende Subjektive Untersuchungen zur Akustik von Konzertsälen Vergleich objektiver Akustischer Parameter mit Ergebnissen subjektiver Untersuchungen an Konzertsälen,' DAGA-Tagung (Aachen 1973), Berichtsheft VDI, Düsseldorf, p. 419.
21. Lehmann, P., (1972). 'Auswertungen von Impulsmessungen in verschiedenen Konzertsälen,' DAGA-Tagung (Stuttgart 1972), Berichtsheft VDI, Düsseldorf, p. 162.
22. Cooley, W.E. and Lones, P.R., (1962). *Multivariate procedure for the Behavioural Science*, John Wiley, New York and London.
23. Cremer, L., (1961). *Die wissenschaftlichen Grundlagen der Raumakustik*, Vol. 2, S. Hirzel, Stuttgart, p. 248.

Index

Absorbers
  flexible frame, 92
  location, 101, 105, 113
  membrane, 92
  mineral wool, 138
  porous, 91, 94, 96
  uneven, 185
Absorption, 33, 38, 51, 78, 87, 90, 108
  air, 87–88, 109, 117, 124, 125
  audience, 157
  coefficient, 15, 87, 91, 95, 96, 101, 106, 107, 112, 113
  effect of paint, 97
  elastic, 93
  molecular, 97, 98
  variable, 54
Acoustic
  aero, 19
  design objectives, 33
  electro, reinforcement, 35–37, 130, 137–144, 165
  feedback, 131
  impedence, 15
  plaster, 14
  psycho, 17, 181
  tile, 14
Air
  conductivity, 97
  dryer, 74, 109
  viscosity, 97
Allan, W, 7
Ambiophony, 171
Amphitheatre, 145, 161–168
Amsterdam Concertgebouw, 44, 47, 50
Ando, Y, 196
Anechoic chamber, 190, 191, 197, 201
Anti-node, 172
Artificial head, 80–81, 200, 210, 218
Arts Council (U.K.), 163

Assisted resonance, 131, 169–179
Atal, B, 102, 197, 200
A.T.V. studios Wembley (U.K.), 166
Auditorium synthesis, 37, 42

Baffles, 60
Bagenal, H, 6, 15, 123, 171
Balcony, 27, 46, 54, 63, 65, 135, 136, 152, 154
Basel, Stadt Casino, 44, 47–49
Bayreuth Opera House, 26, 151, 152
B.B.C., 73–74, 76, 81, 170, 201
Bekesy, 89, 149
Beranek, 213
Berlin Broadcasting SFB, 157
Bipolar rating scale, 215–224
Blavert, 198
Blend, 44
Bolt, R, 1, 3, 7, 13–20, 122
Boston Symphony Hall, 14, 44–46
Boxes, 62, 70
Braunschweig, The Stadthalle, 134–136
Brebeck, D, 108
Bristol, Colston Hall, 45
Britton, B, 6
Broadcasting studio, 74, 130
Broadhurst, 171
Brown, A, 2–3, 7, 161–168
Buenos Aires, Teatre Colon, 45
Building Research Establishment, 6
Burd, A, 73–86, 219

Canac, 148
Ceiling, 60–63, 133, 135, 136
  variable acoustic, 130
Central Hall, York University, 178–179

Chamber Music Hall, 63, 132
Chorus, 54
Cinema, 63
Clarity, 27, 33
Clearness, 224
Clouds, 181
Coherence, inter-aural, 201, 204, 205
Cohesion performance, 33
Computer
  analysis, 78
  simulation, 80–81, 101, 106
Concert hall, 24, 43–44, 47, 51–57, 60–69, 206, 210, 213, 216
Congress Building, Monte Carlo, 159
Correlation
  coefficient, 130, 184, 203, 204
  inter-aural, 130
Correlators, 212
Coupling
  resistive, 92
  stage/hall, 35–36
Cremer, L, 145–159, 224
Cross-talk compensation filters, 200

Damaske, 197
Day, B, 87–99
Decay curves, 140
Definition, 224
Delphi, 10
Delsasso, L, 119, 122, 126
Diffusing elements, 113
Douglas, H, 219
Downes, R, 6
Drama theatre, 63

Early decay time, 68–69, 71, 169
Echo delay time, 193
Echoes, 13, 27, 33, 44, 47, 133
Echograms, 27, 34–35, 38, 42, 149
Edinburgh Opera House, 2–3, 6–7, 85, 115
Effectiveness factors, 28, 32
Energy, 79, 103
  translational, 98
  vibratory, 98
Ensemble, 44

Epidaurus, 145–151
Eyring, 13, 101–113

Fibrous mat, 91–92
Flanking, structure-borne, 77
Fletcher, H, 13, 119, 123
Flow resistance, 91, 94, 96
Fly Tower, 115–118
Ford, R, 92
Fourier transform, 182, 198, 199, 207, 211
Foweather, 171
Franssen, 172
Frequency
  angular, 91
  Napier, 98
  natural, 93, 191
Fresnel's integral, 191

Gerlach, R, 103–113
Glasgow, St. Andrews Hall, 45
Glass-fibre, 190
Gottlob, 197–201, 213
Green, G, 181
Guelke, 171

Harris, C, 116, 122, 123, 154
Harris, S, 8, 83
Hass effect, 191
Hawkes, R, 219
Headphone listening, 80–81
Helmholtz resonator, 94, 173
Herodes Atticus, 147–148

Impedance, complex, 87, 90, 91
Impulse response, 78–80, 109, 117, 182, 184, 185, 190, 199
Incidence, angle of, 15, 90, 91, 104, 112, 147, 181, 184, 185, 190, 191
Ingard, K, 184
Ingerslev, F, 116, 174
Initial time delay gap, 44
Intelligibility, 13, 15, 33, 129, 130, 137, 203

Index

Interference pattern method, 182, 183, 188, 190
Inversion index, 68–71
I.S.O., 95, 96

Jahrhunderthalle Hoechst, 137–141
Jones, 171
Jordan, V, 55–72, 154, 169, 174

Karajan, H. v, 216
Kirchhoff, 124
Kirkegaard, R, 43–54
Kneser's theory, 125
Knudsen, V, 2, 13, 116, 119–127
  Jones audiometer, 122
Kuhl, W, 105, 203, 213
Kurer, R, 149
Kurfurstliche Oper, Munich, 152, 153
Kuttruff, H, 106, 117, 129–144, 170

La Chaux-de-Fonds, Salle Musica, 45
Lambert's cosine law, 104, 106
Leipzig, Neues Genordhaus, 45
Leith Town Hall, 115–118
Leonard, R, 122
Leverkusen, The Forum at, 131–133
Lord, P, 78
Loudness, 35, 137, 213, 214
Loudspeakers, 39–40, 66, 79, 87, 136, 140, 141, 149, 159, 173–179, 191, 214
  directional, 171
  listening, 80–81, 197–206
Louvre surface, 188–190

Maekawa, Z, 181–196
Markoff chain, 101–113
Mass spring system, 92
  law, 166
Matthew, Sir Robt., 2–11, 171
McCormick, M, 92
Melbourne, Victoria Arts Centre, 53
Mellert, 197
Meyer, E, 197, 203
Microphone, 39–40, 51, 63, 71, 75, 80–82, 108, 149, 173–179

Microphone—*contd.*
  close, 170
  directional, 171
  position of, 110, 111
Modelling
  cost of, 83
  design stages, 82–83
  impedance, 89
  mathematical, 78–79
  optical, 27, 77
  physical, 7, 16, 35, 59–60, 62–63, 65, 69–71, 73–79, 81–82, 87, 95, 108, 117, 126, 133
Monk, R, 98
Morse, P, 15
Mozart Hall, Stuttgart, 157
Muller, H, 153
Multi-purpose halls, 23, 35, 43, 52–53, 126, 129–143
Muncey, R, 89
Munson, W, 122
Music
  non-reverberant, 73, 79
  simulated, 87

National Theatre, Munich, 153
New Metropolitan Opera, N.Y., 154
Newman, P, 115–118
Nickson, A, 88
Nitrogen, use of, 109
Noise, 17, 23–24
  background, 51
  criteria, 14, 24
  electrical, 173
  ventilation, 24, 210
Normal modes, 15, 93

Opera House, 43–44, 60, 63–64, 68, 71
Orchestra
  enclosure, 24–26, 40, 51
  studio, 65, 74
Organ, 54
  pipe, 13–14, 40
  transfer, 40
Oscilloscope, 95, 108

Parkin, P, 7, 169–179, 213

Perforated board, 190
Phase angle, 89, 90
Philharmonic Hall, Berlin, 154–156
Philharmonic Hall, N.Y., 181
Phonodrome, 165–168
Photo-electric detectors, 78
Pistol, 7, 13
Plenge, G, 150, 213–225
Porosity, 91
Portland Auditorium, 35
Prinzregenten Theatre, Munich, 152
Proscenium Arch, 25, 40, 46, 116
Pseudo-random maximum-length code 206, 207, 209

Rayleigh, Lord, 124
Recklinghausen Opera House, 163
Recorder, level, 95
   tape, 74, 95, 108
Recording hall, 63
Reflection
   ceiling, 27, 33–34, 41, 155
   coefficient, 102, 181, 188–190
   co-planar, 33
   diffuse, 35, 46, 133
   early, 44, 47, 54, 79, 129
   impulse response, 182
   lateral, 27, 33
   rear wall, 33
   stage, 33
   transfer-function, 182, 185
Reflectors, 54, 60, 69–70
   parabolic, 71
Rehearsal room, 63
Relative humidity, 98, 109, 124
Reverberation, 7, 13–14, 27, 33, 35, 37–38, 44, 51, 54, 65, 66–72, 73, 99, 101–118, 126, 129, 130, 131–144, 154, 169–179, 202, 213, 224
   artificial, 80
   electro-mechanical, 143
   regenerator, 171
   room, 74, 95, 96, 133
Reynolds number, 94
Ripple tank, 77
Royal Festival Hall, 6, 10, 101, 170–179

Rudnick, I, 122

Sabine, W, 13–14, 101, 103–113
Scharoun, H, 155, 157
Schlieren Photography, 77
Schroeder, M, 106, 117, 130, 197–212
Schultz, T, 143–154, 166
Seating, 26–31, 51–52, 54, 60, 67, 147, 155, 215–218
Siebrasse, 197, 201
Singakademie, Berlin, 155
Smoke visualisation, 78
Sommerville, T, 2–3, 7, 213
Sound
   decay, 117, 118
   density, 91
   diffusion, 207
   direct, 37–38, 44, 90
   envelopment, 33–35, 37–38, 169
   isothermal, 96
   rays, 78, 90, 147, 150
   reinforcement, 39–40, 51–52, 54
   reverberant, 35–38
   synthesis of, 80–81
   velocity, 35, 91
Spandock, 73
Spark testing, 7, 63, 65, 108, 201
Speech, 51, 87
Spring, N, 74
Sproson, W, 75, 219
Stage, 23, 26, 33, 35, 46, 51–53, 60–63, 69, 126, 133, 134
   house, 147
Standing waves, 14–15
Stephens, R, 2, 119–127
Stiffness, bending, 92, 93
   cavity, 93
Stokes, 124
Structure factor, 91
Subjective appraisal, 73–74, 80, 87, 130, 215–224
Sydney Opera House, 55, 69–70

Tanglewood Music Shed, 45
Thiele, 203
Timbre, 190, 191

Time delays 38, 80–81
Torrance, V, 1–2
Transducers, 87, 88
Transition probability, 104
Transmission loss, 14

Ultrasonic waves, 77
Utzon, J, 55, 57, 59

Veneklasen, P, 21–42, 122
Vienna, Musikvereinssaal, 44, 47–48

Vineyard-steps, 156

Washington University, Auditorium, 32
Watson, N, 122
Watson, R, 122
Wedge-walls, 158
Wilkens, H, 213–225
Williams, Sir Ralph, 6
Wood, A, 123

Young's modulus, 92, 93